Susanne Meyer
Informal Modes of Governance
in Customer Producer Relations

MEGACITIES AND GLOBAL CHANGE

MEGASTÄDTE UND GLOBALER WANDEL

herausgegeben von

Frauke Kraas, Jost Heintzenberg, Peter Herrle und Volker Kreibich

----

Band 1

Susanne Meyer

Informal Modes of Governance in Customer Producer Relations

The Electronic Industry
in the Greater Pearl River Delta (China)

 Franz Steiner Verlag

Umschlagabbildung: Hongkong
© Susanne Meyer

Bibliografische Information der Deutschen Nationalbibliothek:
Die Deutsche Nationalbibliothek verzeichnet diese Publikation in der Deutschen
Nationalbibliografie; detaillierte bibliografische Daten sind im Internet über
<http://dnb.d-nb.de> abrufbar.

© 2011 Franz Steiner Verlag, Stuttgart
Druck: AZ Druck und Datentechnik, Kempten
Gedruckt auf säurefreiem, alterungsbeständigem Papier.
Printed in Germany.
ISBN 978-3-515-09849-6

CONTENTS

LIST OF FIGURES

LIST OF TABLES

ABBREVIATION

ASEAN	Association of Southeast Asian Nations
CAGR	Compound annual growth rate
CEPA	Closer Economic Partnership Arrangement
CM	Contract Manufacturer
CN	China
CS	Component Suppliers
DC	Direct Contracting
DFG	German Research Foundation
DG	Dongguan
EC	Equity cooperation
FA	Framework Agreement
FDI	Foreign Direct Investment
FIE	Foreign Invested Enterprise
GDP	Gross Domestic Product
GIO	Gross Industrial Output Value
GOL	Generalised Ordered Logit
GPRD	Greater Pearl River Delta
GR	Growth rate
GZ	Guangzhou
HK	Hong Kong
IPR	Intellectual Property Right
JV	Joint Venture
LRM	Linear Regression Model
ML	Maximum Likelihood
NDS	Net Differential Shift
NEC	Non-equity cooperation
NIE	New Institutional Economics
NPS	Net Proportionality Shift
OBM	Original Brand Manufacturer
ODM	Original Design Manufacturer
OEM	Original Equipment Manufacturer

OLM	Ordered Logit Model
OLS	Ordinary Least Square
pp	Percentage points
PRA	Parallel regression assumption
PRD	Pearl River Delta Economic Zone
R&D	Research and Development
RMB	Renminbi
RN	Renegotiation
RVA	Revealed Comparative Advantage
S.D.	Standard Deviation
S.E.	Standard Error
SEZ	Special Economic Zone
SME	Small and Medium-sized Enterprise
SOE	State owned enterprise
SPP	Priority Programme
SSRC	Social Science Research Centre
SZ	Shenzhen
TDC	Hong Kong Trade Development Council
TNS	Total Net Shift
US	United States
USD	United States Dollar
VC	Value Chain
VAT	Value Added Tax
WTO	World Trade Organisation

1. INTRODUCTION

1.1. BACKGROUND INFORMATION AND NECESSITY OF RESEARCH

The global impact of China as one of the world's leading production engines is growing. An increasing number of goods on the world market are produced in China. The Greater Pearl River Delta (GPRD), consisting of Hong Kong (HK) and the Pearl River Delta Economic Zone (PRD), part of the southern Chinese province of Guangdong, is to a large extent responsible for the economic growth and increasing prosperity of China. Since the implementation of the Chinese opening policy in 1979, the country has attracted foreign investments from HK and, more recently, from Taiwanese and Western firms. Due to its proximity to HK, the PRD served as an experimental zone for Chinese policy makers to coordinate foreign interests. Low production costs were the initial motivation for foreign firms to invest in or source from the GPRD. Today, the PRD is one of China's most important economic centres. The economic power of the GPRD results from the intensive interaction between HK and the PRD. Meanwhile, the GPRD is exposed to intense competition from other countries and regions which also offer attractive production conditions for global firms. Moreover, production costs in the GPRD are rising due to higher wages, higher land prices, higher electricity costs, etc. Although some firms have relocated operations to other countries, the majority of firms continue to source or set up production facilities in the GPRD. The trend has decelerated as a result of the economic crisis in 2008, but is still obvious. Firms in the GPRD seem to be able to maintain their competitiveness on the world market despite severe competition and higher production costs (FEDERATION OF HONG KONG INDUSTRIES (FHKI) 2007; FEDERATION OF HONG KONG INDUSTRIES (FHKI) 2003; ENRIGHT et al. 2005).

The new challenge of serving leading global firms provides explanations. In times of made-to-order production, shorter product life cycles, smaller batch sizes, high market volatility and especially mass production, a flexible firm and network organisation is required in order to provide speed in the production process and the necessary ability to rearrange the production process (VOLBERDA 1996). Besides cost cutting and upgrading pressure, the *flexibility of firms* is crucial for their integration into global value chains. The model of flexible specialisation and the just-in-time concept addressed this new global challenge for the first time. Although some research exists about the means of achieving flexibility (for example small-batch production, economies of scope, flexible machines, specialisation), the contribution of informal modes in interactions with business partners to smooth relations is still underrepresented in the economic geography literature. This work aims to explain how firms in the GPRD have adapted their business organisation, especially their customer and producer relations, in order to build a flexible production network which is capable of ensuring their competitive advantage.

The concept of this work acknowledges the call of YEUNG (2007) and YEUNG and LIN (2003) for a research agenda on Asian business that integrates micro-level analysis of intra-firm governance and macro-level influences of institutional settings as well as global production networks. Existing literature studies the growth and success of firms in China and especially of the firms in the PRD (ENRIGHT et al. 2005; BERGER and LESTER 1997). Moreover, Asia-specific literature provides insights into the peculiarities of Asian businesses, such as informal institutions in guanxi networks (GIPOULOUX 2000; XIN and PEARCE 1996; ZHANG and ZHANG 2006).

The informal ties which ease and smooth relations between business partners seem to play a significant role in firm organisation. The concept of informalisation does not refer to illegality or the informal sector. It goes beyond that and touches on the personal elements of interactions which allow the avoidance of time-consuming official business procedures. This shortens firms' time to market and provides an additional opportunity for quick response which increases firms' flexibility. Until now there has been no comprehensive study of the informal way Asian firms do business and the effects of these methods on the flexibility of firms and their success. The aim of this work is to shed light on that issue, thereby filling an important gap in research.

Informal modes of interactions seem to characterise both the external and the internal organisation of firms. The competitive advantage of the GPRD is expected to result not only from a single firm's capability to organise and react flexibly to the needs of global customers, but also from the way a flexible network of firms is built. This work concentrates, therefore, on the organisation of customer producer relations along the value chain in the GPRD. Emphasis is put on the unique institutional setting in the GPRD. HK provides a legal system adapted to fit international standards, whereas firms in the PRD operate in a transitional setting.

This work could potentially be developed further within the Priority Programme 1233 "Megacity—Megachallenge: Informal Dynamics of Global Change", funded by the German Research Foundation (DFG). This would provide an opportunity to study the relevance and significance of informal modes in customer and producer relations of firms in the GPRD in order to enhance the flexibility of firms. Informal modes of governance in customer and producer relations cannot be seen as the only factor influencing flexible firm organisation, but rather as one of several possibilities. The GPRD was selected as a research region because its economic structure requires continuous adaptation to changes in global markets and to the institutional setting.

A sectoral study of the electronics industry was conducted. The novelty of the research topic suggested a concentration on a specific sector in order to exclude the influences of industry specifications. The electronics industry is the most relevant and dynamic in the GPRD. It serves as a good example of the interplay between informality and flexibility. Two company surveys were conducted in HK and the PRD, complemented by qualitative interviews with regional stakeholders.

1.2. RESEARCH OBJECTIVES AND QUESTIONS

The research objectives of this work are to discover:

1. how electronics firms in the GPRD are spatially organised and integrated into global value chains and what pressures for flexibility driven by global markets and institutional changes they have to respond to

2. which governance modes HK and PRD firms apply in customer and producer relations along the value chain and how they can be explained

3. how relevant informal modes of interactions between customers and producers are in achieving a high level of flexibility under special consideration of:

   • contact and selection processes for customers and producers

   • contractual arrangements

   • enforcement mechanisms of contracts

The main research questions addressed in this work are:

Theory-guided

A Which critical factors pushed the global spread of production locations, especially in the electronics industry? How does this affect the organisation of value chains? Which new challenges and risks do electronics firms in emerging economies face if they want to enter the global production network?

B How does the institutional environment in a region determine the behaviour of firms under special consideration of institutional change in transitional economies?

C How does the institutional environment and the transactional specificity affect the choice of governance modes? What governance modes can be expected in HK and the PRD?

D Under what circumstances do firms prefer informal practices over formal ones to organise their customer producer relations? How do informal interactions enhance the flexibility of firms? In answering these questions, the focus will be on:

- contact and selection procedures for customer and producers
- contractual arrangements
- enforcement mechanisms

Methods

E How can informality and flexibility be operationalised in customer and producer relations?

F Which methods are suitable for measuring informality and flexibility in customer producer relations?

Empirically-guided

G How are electronics firms in HK and the PRD integrated into global value chains? How is the division of work spatially organised between HK and the PRD?

H How well is the formal institutional environment in HK and the PRD developed? What informal institutions guide the way of doing business in HK and the PRD?

I Do HK firms organise their business relations to customers and producers in the PRD differently from their relations to customers and producers abroad? Is this a consequence of the different development stages of the institutional environment? Is there a trend toward hybrid forms of governance in China resulting from the transitional change?

J How high is the degree of informality in customer producer relations of HK-based and Chinese-based firms? Are there any differences in the behaviour of firms between the two locations? Does informal behaviour result in greater flexibility? In answering these questions, the focus will be on:

- contact and selection procedures for customer and producers
- contractual arrangements
- enforcement mechanisms

Policy-guided

K What implications do the research results have for policy makers and firm managers in the GPRD and what further research is needed in order to consolidate the new business concept?

1.3. STRUCTURE

The main body of this work has been divided into four parts. The first part deals with the conceptual framework (Chapter 2), the second part with operationalisation, methods and data (Chapter 3), the third part (Chapter 4, 5 and 6) addresses the empirical findings of the work, and finally, a conclusion is drawn which includes policy recommendations and further research implications (Chapter 7).

The conceptual Chapter 2 discusses how firms in emerging markets can successfully integrate themselves into global value chains under special consideration of leading Western firms' requirements for flexibility. In doing so, the focus is placed on how firms organise their customer producer relations in order to build a flexibly organised network of firms along the value chain. To answer the theoretically guided questions **A—D**, an eclectic concept is applied. The behaviour of firms is discussed against the background and interplay of the industrial dynamics and market conditions, the location of specific formal and informal institutions and firms' capabilities to deal with them. Firstly, the industrial and market dynamics implying the need for firms' flexibility are analysed using the global value chain concept of GEREFFI et al. Secondly, a discussion about the changing institutional environment in transitional economies and the interplay of formal and informal institutions guiding firms' behaviour is outlined. Thirdly, the consequences for the selection of governance modes to organise customer and producer relations will be examined. Finally, the discussion goes beyond governance modes and deals with the expected informal modes of behaviour applied in contacting and selecting customers and producers as well as in concluding and enforcing contracts. All four conceptual discussions are based on literature reviews and conclude by producing hypotheses which will be proven in the empirical part of this work.

Before analysing the empirical data, Chapter 3 describes how informal modes of interaction between customers and producers and increasing firm flexibility can be operationalised and measured using primary and secondary data and literature in order to answer research questions **E—F**. Because of the novelty of the topic (informality achieving flexibility), emphasis is placed on the collection of primary data. A standardised survey of electronics firms was conducted in HK and another in the PRD. Additionally, interviews with major players in the electronics industry in HK were organised. The collection of empirical data was supplemented by expert interviews in order to gain a deeper understanding of the processes in the research region. This chapter also covers the assessment of research methods and a discussion about data quality and validity.

The first empirical Chapter 4 analyses the framework conditions influencing the behaviour of electronics firms in the GPRD. Firstly, the industrial dynamics of the electronics value chain are analysed through the use of primary and secondary data. Secondly, the changes to the formal and informal institutional environments in China are studied. The second empirical Chapter 5 investigates the applied governance modes of HK and PRD firms for their customers and producers as a consequence of the industrial dynamics of the electronics industry and the institutional environment in HK and China. Some governance modes are assumed to be more informal and flexible than others. Two logit models are built to identify the determinants of governance modes in order to prove or disprove respectively what has been discussed in the conceptual part of the work. The third empirical Chapter 6 goes beyond the governance modes and addresses in more detail how relevant informal modes of interactions with customers and producers are in terms of contacting and selecting, contractual arrangement and enforcement of contracts. Attention will be paid to differences in the organisation of firms in HK and the PRD in consideration of the distinct institutional environments in which they operate. After every empirical chapter, a *summary and discussion section* is included to refer to the theoretically guided hypotheses which provide the background for answering research questions **G—J**. Mirroring the empirical data with the conceptual discussion seems to be valuable in terms of providing new inputs for the refinement of theoretical concepts. Those concepts are mainly derived from experience with Western firms and tend to underemphasise the specific way in which Asian firms organise their business.

Finally, in Chapter 7 a conclusion is drawn for policy implications (research question **K**). As this work is based on fundamental research of an explorative nature, further research topics are recommended in order to improve the academic understanding of the interplay between informality and flexibility.

2. DEVELOPING A CONCEPTUAL FRAMEWORK: ORGANISATION OF CUSTOMER AND PRODUCER RELATIONS IN EMERGING MARKETS

This chapter aims to give a theoretically supported background to how firms are expected to organise their customer producer relations. In the Section 2.1, a model developed by LEWIN et al. (1999) will provide a framework illustrating that external determinants such as institutional and industrial dynamics coevolve with a firm's organisation of its customer producer relations. The structure of this chapter is developed according to this model. The use of informal interactions as a contribution towards enhancing the flexibility and success of firms is critically assessed. HERRIGEL (2007:2) summarised as follows:

> "It was clear that in order to understand the success of the alternative more flexible forms of organization [...], one had to look past the boundaries of the firm and see how producers were embedded in regionally specific institutions and networks."

2.1. COEVOLUTION OF FIRMS' BEHAVIOUR, INDUSTRIAL DYNAMICS AND INSTITUTIONS

In this section, emphasis is placed on how the behaviour of firms can be analysed according to recent management literature. A coevolutionary perspective suggests concentrating not only on how firms organise their business activities, especially their customer and producer relations, but also on environmental factors influencing the organisational form. The theoretical model of organisational adaptation by LEWIN et al. (1999) considers the interaction of the firm, its industrial dynamics and its institutional environment. The basic thesis of this model is that a firm's adaptation processes and its organisational form coevolve with changes in the institutional environment and the industry. Industrial changes caused by the introduction of new technology, for example, require greater institutional protection, which leads to new organisational forms of firms and the weakening of existing ones. If the formal institutional environment is incomplete, firms attempt to adjust by expanding the firm's network, intensifying relationships to customers and producers or diversifying their range of partners to maintain their competitiveness. The model of LEWIN et al. (1999:235) attempts to integrate the interplay between "adaptation of individual firms, their competitive dynamics and the dynamics of the institutional systems within which firms and industries are embedded". PAJUNEN and MAUNULA (2008) extended the model by emphasising its importance

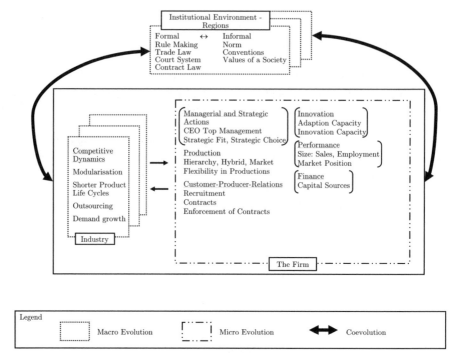

Figure 2.1.: Model of organisational adaptation

Source: Adaptation to LEWIN et al. (1999:537)

for internationalisation processes. They strengthened the influence of firms' own organisational resources and capabilities and their ability to analyse the existing and emerging industry constraints, especially in an international environment. Moreover, KRUG and HENDRISCHKE (2008) proposed a new institutional perspective to explain the local differences in business systems in China. They suggested studying institutions on a regional rather than a national level, as done by LEWIN et al. (1999). With respect to the three papers, this work focuses on the coevolution of the global industrial dynamics, the regional institutional environment and the firms' adaptation processes to successfully deal with them. Figure 2.1 provides a framework for this study to analyse how firms organise their business activities and why they do it in a particular way. The increasing turbulence of firms' environments usually results from *dynamics of industries*. Firms have to adapt and develop their skills to encounter industrial changes. Especially in a globalised world, firms in developing countries and emerging markets are greatly affected by global changes in their industry. The industrial dynamics can be observed on the macro level, e. g. the global level. The institutional environment influencing the behaviour of firms can be divided into *formal and informal institutions* on the regional level (meso level). Formal institutions are developed on a regional level, such as the rules of law, capital markets, the rule-making system, the education system etc. Informal institutions exist in a region's society, for example culture, norms, conventions and values rooted in its history. Institutions are created to buffer economic entities, groups and individuals from the effects of environmental uncertainty (LEWIN et al. 1999:541). An institutional system can increase or decrease the attractiveness of regions for business, because it is an essential factor for the decision-making process and behaviour of firms. The specific institutional system enables and restricts the adaptation and development of the governance modes of firms. Institutions and industrial dynamics influence *the organisation of firms* on the micro level. Managerial and strategic actions, financial sources, production and innovation organisation and customer and producer relations are determined by the industrial dynamics and the institutional environment. But it is not a one-way determination. The organisation of firms also influences the institutional environment and the industrial dynamics. It is a mutual interaction. Moreover, industrial dynamics coevolve with the regional institutional environment.

Research into a firm's organisational pattern for customer and producer relations should not be limited to a firm's level analysis, but should be extended to the study of the interrelation of industrial dynamics and the institutional environment, as recommended by LEWIN et al. (1999). Therefore, Section 2.2 focuses on the *development of industries* at a global level. Emphasis is placed on the challenges firms in emerging economies encounter while attempting to integrate themselves into global value chains. It will become clear that achieving flexibility is the major challenge for firms involved in the manufacturing process. In Section 2.3, the *formal and informal institutions* determining a firm's decision-making on customer and producer relations will be analysed. Additionally, the subject of institutional transformation will be outlined in the context of China's transformation from a planned to a market economy. In HK, however, a market-institutional environment exists. The influence of industrial dynamics and the institutional environment on *firms' governance mode of customer producer relations* will be studied in Section 2.4. Section 2.5 will investigate under which conditions informal or formal aspects of behaviour contribute to the financial success of firms. Here, emphasis is placed on recruitment methods, contractual arrangements and enforcement mechanisms of contracts. The need for informality in business relationships is discussed against the background of the existing industrial and institutional environment.

2.2. THE NEED FOR FLEXIBILITY IN EMERGING ECONOMIES: GLOBAL VALUE CHAIN PERSPECTIVE

This sections aim to explain how the electronics industry develops globally and how this influences the behaviour and adaptation processes of firms in emerging economies (see model LEWIN et al. (1999)). Focussing on the global development of industries, the empirically-based concept of modular value chains provides a framework to analyse the competitive dynamics of the electronics industry. The approach sheds light on new challenges encountered by firms at the upper end of the value chain. It will become obvious that the cost-effectiveness of firms in emerging economies is not the only competitive advantage, but also the flexibility to adapt to changing customer requirements, which has increased in importance recently (see Section 2.2.1). The many facets of flexibility will be discussed in Section 2.2.2.

2.2.1. Global Production in Modular Value Chains

The globalisation of production and markets has led to the integration of firms in developing countries and newly industrialising countries into the global economy (ERNST 2004). The electronics industry is characterised by a rapid fragmentation (LALL et al. 2004; JONES et al. 2005) and an increasing trade in components across the globe (WHITFORD and POTTER 2007). However, for the analysis of the shape and trajectory of global economic integration, a tool is needed which not only focuses on the macro-level statistics, such as trade and investment, but also identifies the players, links and locations in order to make detailed contours of the world economy visible (STURGEON 2000:1–6). STURGEON (2002, 2003a, 2006) and GEREFFI (2005) developed the concept of modularity in value chains to conceptually describe the empirically observed restructuring of production organisation in the electronics industry. The concept stresses the spatial scale and the productive players. GEREFFI et al. (2005:97) states that "the modular form [of the value chain] appears to be playing an increasingly central role in the global economy", as standards, information technology and the competencies of producers improve. The modular value chain concept provides a suitable framework to answer the first theoretical question. It explains:

- which critical factors pushed the global spread of production locations, especially in the electronics industry

- how this affects the organisation of value chains

- which new challenges and risks firms of emerging economies face if they want to integrate themselves into the global production network.

Concept of Modularity in Value Chains

In a modular chain, transnational corporations (leading firms) outsource production activities to large suppliers (contract manufacturers). GEREFFI et al. (2005) characterise *leading firms (LFs)* as those that usually initiate the flow of new products through the value chain and help to drive the organisation and geography of their production networks by demanding their producers engage in new activities and invest in new places. In the electronics industry, LFs include firms such as Dell or IBM. *Contract manufacturers (CMs)* provide a full range of services without a great deal of input from LFs. LFs provide only instructions as to WHAT to make, but it is almost entirely up to the suppliers HOW and sometimes WHERE this is done. CMs take over complex production processes from LFs and have their own research and development facilities. CMs usually serve more than one LF to economise on scale production (see Figure 2.2). Therefore, they use generic machinery that limits transaction-specific investments, but supports spreading investments across a wide customer base. Suppliers provide "modular" production for their customers. In China, CMs in the electronics industry are also known as *Original Equipment Manufacturers (OEM)* or *Electronic Manufacturing Services (EMS)*. Originally, the term OEM was used for LFs which manufactured a significant proportion of their products, but later, with the shift of production to CMs, the term OEM shifted as well. In contrast, the automobile industry still reserves the term OEM for the large brands such as Ford or Toyota, because they still provide parts and subassembly lines. The terminology of suppliers in the automobile industry is well known—there are first (also system suppliers), second or lower tier suppliers. In the electronics industry, CMs correspond to the first tier suppliers and *component suppliers (CSs)* to the second or lower tier suppliers. CSs provide CMs with components and services for simple goods, because of their limited capabilities (STURGEON 2000:8–11). The industry-specific terms are different, but they roughly denote the same scope of activities. As this work will investigate the electronics industry, it will refer to the terminology of this industry: leading firms, contract manufacturers and component suppliers. The main drivers for a shift in the electronics industry from a completely in-house, hierarchical value chain to one which is modular-based are multifarious:

1. With the economic boom in industrialised countries in the 1970s, **international competition intensified**.

2. **Market volatility** in the electronics industry has made the scheduling of production extremely difficult. CMs have provided the opportunity to increase or reduce the volume of production at short notice.

3. The **shortened lifecycles** of electronics products have placed pressure on the capacity of firms to face regular changes in all units (R&D, manufacturing,

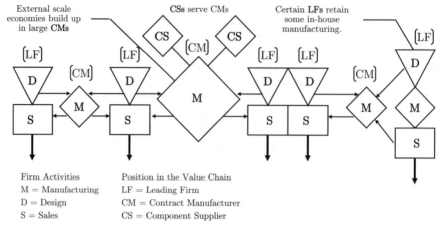

Figure 2.2.: The modular production network pattern

Source: Adaptation to STURGEON (2002:477)

marketing, sales). The division of work has made the shorter product life cycles manageable for every unit.

4. The **complexity of products** has increased, which has led to more steps in the production process and has resulted in the potential for fragmentation of production.

5. The **competence of suppliers** has improved in emerging economies. Firms from industrialised countries are now more willing to divest their manufacturing divisions.

6. The rapid **advance in organisation and information technology** has reduced the difficulties of coordinating production at a distance. The development of computer-controlled production equipment on the factory floor has led to the **codifyability of data** and its exchange. Manufacturing activities and enterprise resource planning systems have been introduced and have tracked the inventory. More recently, supply management systems have been developed which calculate the demand forecast for component purchasing, inventory management and capacity planning.

7. Expansive **automation** in the manufacturing process requires high volume production. As markets are volatile, LFs face the risk of low plant utilisation. It has been more efficient to pool capacity in shared external suppliers.

8. The ascendancy of **liberalisation policies** around the world has made the trade between nations possible or easier to manage.

9. **Standards** have been established (for example the International Organisation of Standardisation (ISO), International Electrotechnical Commission (IEC)), which have made the understanding of product specification easier by reducing the amount of information which needs to be exchanged.

10. In the beginning, CMs sourced components from their customers. After CMs were successfully established in the mid-1980s, they began to **purchase components from third-party component suppliers**, which made it beneficial for CMs to have locations closer to the CSs mainly located in emerging economies.

11. CMs have **added a variety of front- and back-end services** (design, testing, packaging, after-sales service), which has made them more attractive to LFs (STURGEON 2002:462; WHITFORD and POTTER 2007:12; VOLBERDA 1996:359; ERNST 2004:90; STURGEON 2007).

The interplay between the different development processes resulted in the establishment of CMs which serve a broad customer base. CMs thus had the opportunity of simply applying more of the production capacity to customers who had gained market shares, while reducing the production for firms that had lost market shares. Resource pooling has led to a better plant utilisation and scale economies in equipment and component purchasing. Economies of scope have also been realised through a better asset utilisation, as human effort and fixed capital can be pooled, conserved and deployed across a broad customer base. A specialised firm producing a particular component for several LFs has longer production runs than any LF. Shared producers reduce production costs (MCMILLAN 1995:206). Dynamic advantages refer to the upside and downside flexibility of CMs. They provide LFs with the opportunity to increase or decrease production volume without installing in-house capacity.

The first CMs were established in North America (for example Solectron), but with the increasing intensity of competition and competencies of component suppliers in emerging economies, CMs have shifted their manufacturing activities to low-cost production countries in proximity to their new CSs (ERNST 2004:93). Because of the professional cooperation between LFs and CMs, spatial proximity is no longer necessary, with the exception of some CMs which resist codification. Even though ties are thin, LFs and CMs depend greatly on each other (WHITFORD and POTTER 2007:16). In contrast, the propinquity to Asian CSs is even more important, as the relationships are built less on professional codified contracts and more on relational aspects. Moreover, proximity between plants of the CMs and CSs brings the advantage of component delivery on short notice. Besides geographical proximity, cultural and social proximity have a positive effect on efficient customer producer relations (GOFFIN et al. 2006:193). Besides the shift from Western CMs to Asia, Asian-based CMs (for example Hon Hai Precision (Taiwan), Wong's Electronics (HK)) emerged in first generation countries (Japan, Taiwan, South Korea, HK, Singapore) and later in second generation countries such as China, Vietnam or Thailand. They had a better connection to local CSs (STURGEON 2007; STURGEON 2002; ERNST 2004; STURGEON 2006a). Although Asian-based CMs have fewer globally-spread networks than CMs from developed countries, their global share is growing (ERNST 2004:99; REED BUSINESS INFORMATION n. d.). Component sourcing and actual production activities shifted quickly from emerging countries of the first generation to countries of the second generation.

The spatial proximity of CMs and their CSs results in huge firm agglomerations (STURGEON 2003:216). One should be aware that new industrial agglomerations in emerging markets do not occur because of the spatial dispersion of production, but because of the spatial fragmentation of production. A glance at Asia shows that the trade pattern is predominantly inter-firm, rather than intra-firm. Studies focussing on China have empirically confirmed the emergence of industrial agglomerations, particularly in the Pearl and Yangtze Delta region (FAN and SCOTT 2003; CHRISTERSON and LEVER-TRACY 1997). This clustering of firms is associated with higher productivity. Face-to-face contacts, reciprocity and mutual understanding are crucial for business success in Asia (WHITFORD and POTTER 2007:14). Even if a codification of information is possible, the institutional environment does not sufficiently protect and support codified communication. Therefore, ties between these firms are expected to be not thin and codified, but rather thick (WHITFORD and POTTER 2007:16). STURGEON (2002:488) summerised the performance advantages of modular value chains as follows:

> "First, geographic flexibility creates greater access to a variety of place-specific factors and markets. An important result of geographic flexibility is the system's easy reach into areas with lower factor costs. Second, shared suppliers can better match and adjust capacity to demand, resulting in more intensive capacity utilization. The overall result is lower costs and less risk [...]."

Breaks in the chain tend to emerge where information regarding product and process specifications can be highly formalised (see Figure 2.3). Within individual firms, information creation and exchange is based on tacit links, but between firms, information is codified in the form of protocols and standards (STURGEON 2003:221). Codified links allow the transfer of detailed product and production specifications to occur more easily at great distance. The LF continues to carry out such firm activities as development of production strategy, R&D, design for manufacturing, core design, prototype fabrication, administration and later marketing, sales and distribution. CMs are responsible for process R&D, purchasing components, manufacturing, testing and packaging. CSs provide manufacturing capacity for CMs. The modular concept of value chains expects links to be highly formalised. This implies little time to establish and re-establish inter-firm links. But although this formali-

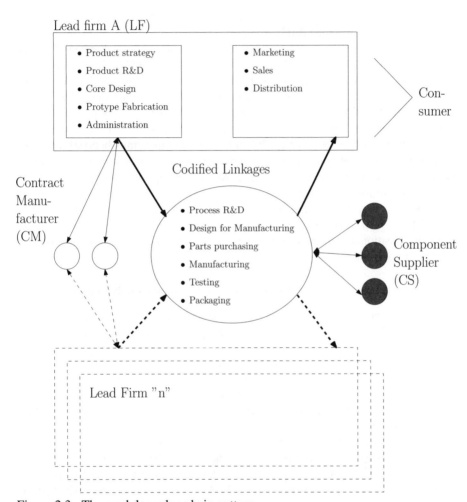

Figure 2.3.: The modular value chain pattern

Source: own draft according to STURGEON (2006a) and STURGEON (2002)

sation is assumed, HOWARD and SQUIRE (2007) provide empirical evidence show-
ing that the modularisation of production leads to greater collaboration between
customers and producers and reduces the probability of working in "arm's length"
market relationships.

This work mainly concentrates on the upper end of the value chain—the link between CMs and CSs in the GPRD. The aim is to make a comparison to ties between
LFs and CMs. Empirical evidence for the formalisation of ties between LFs and
CMs exists, but the characteristics of ties between CM and CSs have received less
attention in the past. Even though industrial dynamics are pushing the codification
of information, the institutional environment, for example in China, acts restric-
tively on this issue. When the transfer of data or information is not protected, firms
still have to rely on links based on proximity and informal ties. It takes some time
to make the links work efficiently. It can be expected that links between CMs and
CSs in emerging economies with an incomplete formal environment rely heavily on
proximity and less codified cooperation. As the concept of modular value chains
lacks an institutional perspective, it did not consider this aspect. This work will fill
the gap by comparatively studying the locational and organisational pattern of LFs
and CMs, as well as that of CMs and CSs. Differences between the behaviour of
firms in transitional countries such as China and in market-led economies such as
HK should become visible.

Challenges and Risks of Firms while Acting on the Global Platform

The intensive outsourcing of production activities from LFs to globally acting and
spreading CMs has raised *challenges as well as risks for the LFs*. They have been
able to overcome the risk of high market volatility by outsourcing production activities. LFs thus had the opportunity to alter the volume of production upwards
and downwards at very short notice without installing any of their own plants and
equipment, which saved them costs. The LFs concentrate on the strategic and organisational leadership beyond the resources. They retain in-house activities in which
they have a particular strategic advantage, while outsourcing those in which they
are not competitive. LFs offer technological and quality-control advice to suppliers and give extra design responsibility to them. Suppliers are monitored and are
requested to deliver inventory just-in-time (STURGEON 2002:455; HENDERSON
et al. 2002:443; MCMILLAN 1995). But the LFs have to endure the risk of losing
their ability to innovate and design successive product generations, which comes
with the loss of manufacturing knowledge. Intellectual property and other sensitive
information about product design, product features, pricing, production forecasts
and customer information can find their way from the shared suppliers to their com-
petitors. CMs gain a strategic insight into their LFs' business, which could be quite
damaging if leaked to other LFs. Moreover, LFs face the risk that a competitor could
easily purchase the common supply firm. Additionally, they fear that CMs may de-

velop their own products and compete directly with the LFs. A further risk is that an outsourcing strategy goes along with commodification of individual firm functions. As the rate of outsourcing increased in the 1990s, firms found themselves managing an increasing number of relationships. Because of that, firms started to concentrate on fewer suppliers with a higher volume (MCMILLAN 1995:203). If the volume is higher, LFs can insist on customised treatment of CMs. When such specific assets are in place, it becomes more difficult for either party to switch. Long-term relationships developed and competitive bidding to select suppliers became less important (STURGEON 2006a:26–42; MCMILLAN 1995).

But nevertheless, LFs are still in the position to select suppliers. Only suppliers which fulfil the LFs' requirements can be integrated into the global production network. Once a supplier is selected, the process of integration advances and the interdependence grows. STURGEON (2002) and ERNST (2004) identified *challenges and risks for CMs and CSs* in emerging economies involved with being selected as a supplier for LFs.

Challenges

- **Reduction of production costs:** The reduction of costs in their own production processes is one of the main reasons why firms from industrialised countries look to Asia. Cost reduction results from decreased location-specific land and labour costs, and decreased expenditure for machinery and components. As the competition between producers in emerging economies grows and LFs look for the best deal, producers continually have to cut costs.

- **Mass Customisation:** Asian producers have to adapt their production processes to a broad base of customers who continually change their requirements. They therefore have to specialise in cross-cutting base processes (processes which support the production of products for a wide range of end-markets), base components (products which can be used in a wide variety of end-products) and base services (services which are needed by a wide variety of end-users). To do so, they have to work with generic manufacturing machines. This allows large product variations as long as the specifications fall within the parameters of the base process and at the same time realise economies of scale. But CMs and CSs fear the risk of focussing too much on standard components, which leads to a loss of distinctiveness in their products. To summarise, their products have to be as standardised as is necessary, and as customised as possible to ensure an adaptable production mode.

- **Speed of reaction:** When Asian producers (CMs and CSs) want to integrate themselves into global production networks, they have to prove their ability to:

- respond at short notice to unexpectedly high demand or a sudden reduction of product volume

- shorten production lead time (quick delivery)

- change product lines and find new CSs quickly, due to the demand for small batch runs

The former weaknesses of LFs are the strengths of Asian suppliers. The dimension of time is one of the most competitive challenges today (REICHHART and HOLWEG 2007:1144–1145).

- **Variety of options:** Asian producers are being pushed hard by LFs to take on additional functions, especially made-to-order final assembly and design services for low-end products. The increased scope of activities and business solutions provides LFs with more options to act and producers with more options to react. Having a variety of options with which to react to changing demand enables CMs and CSs to manage the continual organisational adaptation.

- **Innovation activities and technological capabilities:** The challenge is to keep track of new developments on the one hand while trying to standardise as much as possible on the other hand. For the time being, firms are strongly pressured to build up technological capabilities in order to assimilate new knowledge in design and development. New machinery, R&D labs and qualified workers are essential. Even though not all firms are forced to become highly innovative, the existence of innovation activities will be crucial for survival in the future (STURGEON 2002:455; STURGEON 2007:63).

Risks

- **Reversibility:** CMs and CSs fear the risk that LFs may indeed push for outsourcing, but this process is not irreversible, especially in highly volatile industries.

- **Greater competition:** Another risk is the number of LFs, which declined during the economic downturn in the 1990s. Additionally, the number of CMs and CSs is increasing continually. The position of CMs is weakening and LFs are now much more demanding (for example on price issues).

- **Competing interests:** There seems to be a conflict between the interests of LFs, which are looking for flexibility to scale production up and down, and CMs and CSs, which are looking for predictability and scale (ERNST 2004:104–106).

2.2.2. Flexibility of Firms as a Competitive Advantage in Emerging Economies

The challenges encountered by manufacturing firms in emerging economies when entering the global production network are identified in Section 2.2.1—*reduction of production costs, mass customisation, the necessity for speed and variety of options to react and expansion of innovation activities*. As this work examines the customer and producer relations of firms, only two of the five identified challenges are related to this issue: the *variety of options* for reacting to market changes and the *speed* with which they can be implemented. The adjustment of individual firms to market changes (for example new technology, high volume demand) is not sufficient; the entire value chain has to adapt. If LFs require a certain shipment, CMs and CSs have to work cooperatively in response to short-term dynamics (KRAJEWSKI et al. 2005:453). The successful implementation of production cost reduction, mass customisation and expansion of innovation activities are mainly dominated by the in-house activities of firms. Cost reduction refers to cost-cutting of labour, land, machinery and sourced components. Producers only come into play concerning the price of sourced components. As this work concentrates on the organisation of customer producer relations, prices of compontents are not of great relevance. Mass customisation refers mainly to the assignment of personnel and machinery to changing orders. As the in-house management is responsible for the organisation of processes and services, customers and producers do not influence the plant adaptability. Moreover, the establishment of innovation activities does not depend much on the organisation of custome producer relations either.

VOLBERDA (1996:361–363) looked into the subject of firm flexibility in his highly regarded article "Towards the Flexible Form" and identified speed and variety of managerial capabilities of firms as main drivers for organisational flexibility. For VOLBERDA (1996:361)

> "Flexibility is the degree to which an organization has a variety of managerial capabilities and the speed at which they can be activated, to increase the control capacity of management and improve the controllability of the organization."

VOLBERDA's understanding of flexibility fits particularly well with what capabilities firms in emerging markets require for integration into global value chains. He defines the *variety of managerial capabilities* as a summary of the arsenal of capabilities currently used plus the potential of flexibility-increasing capabilities. A firm must have a variety of capabilities, which increases the variety of options for reacting to different disturbances in the environment. *Speed* in reaction serves as the second essential factor of flexibility. Therefore, *supply flexibility* can be recognised as a firm's ability to respond in multifarious and quick ways to changing demand, using its customer producer network. As the variety of options has a qualitative component and the speed with which actions are carried out has a quantitative component, the following work will use the terms *qualitative* and *quantitative flexibility*

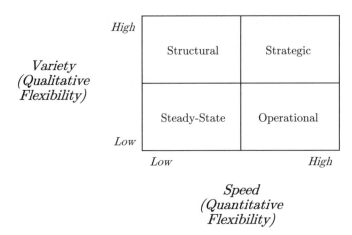

Figure 2.4.: Types of flexibility

Source: According to VOLBERDA (1996:361)

(see Figure 2.4). Two points have to be made: firstly, supply flexibility can be seen as a new challenge for firms in emerging markets integrating themselves into the global economy, and secondly, supply flexibility is highly influenced by the form of collaboration with customers and producers (KRAJEWSKI et al. 2005:455). Supply flexibility is not exclusively responsible for a firm's manufacturing performance, but it makes a contribution. Depending on the organisation of customer producer relations, firms have more or fewer options to adjust to the requirements of LFs. If firms have to increase their production volume at short notice, for example, they can employ additional producers or negotiate and squeeze their existing producers. Speed is necessary to activate and apply a firm's options in a certain time. Firms have to decide which options for responding to higher volume demand take less time. Is the recruitment of a new producer more beneficial than using experienced producers? A considerable number of options for reacting to changing demand, as well as the speed of applying them, depend on the organisation of customer producer relations. They are crucial for the success of the entire manufacturing system (REICHHART and HOLWEG 2007:1148). Relations to customers and producers organised in a way that contributes to speed and variety is assumed to be optimal for a firm's total performance.

    VOLBERDA (1996) distinguished four main types of flexible firm organisation—steady-state *(low variety, low speed)*, structural *(high variety, low speed)*, operational *(low variety, high speed)* and strategic *(high variety, high speed)*. Each type represents a combination of speed and variety. The three main research areas of this study—recruitment of customers and producers, conclusion and enforcement of contracts—can be transferred to the four different types of flexibility. The number of reliable and successfully functioning recruitment processes for customers and producers which firms can revert to indicates the factor of variety. The time firms actually need to employ a new customer or producer is related to the factor of speed. It is expected that firms which consider more options for recruiting partners have a better chance of quickly finishing the process. They would reach the maximum flexibility—strategic flexibility. But there might be firms specialising on a small number of recruitment processes which are very effective. They would refer to operational flexibility. With regard to contract conclusion, firms can have a standard procedure or they can have several contract models. Finally, the time they need to conclude contracts is essential. When it comes to conflicts, firms may take advantage of a large number of different enforcement mechanisms for contracts involving a certain time. Assessing firms' qualitative and quantitative flexibility in the various research areas leads to a classification of firms using the four types of firm flexibility (VOLBERDA 1996:361–363; AVITTATHUR and SWAMIDASS 2007:720; LUMMUS et al. 2003).

LEE (1997) attributed the major economic success of HK firms to their flexibility. He argues that their flexibility arises from the ability of manufacturers to find new market niches and to respond quickly to market changes. Furthermore, he argues that "the flexibility of the Hong Kong economy stems from the interactions between the business strategy of Hong Kong entrepreneurs and the small and medium-sized firm (SME) subcontracting networks" (LEE 1997:189). But there is a distinction between the type of flexibility expected from CMs and CSs in HK and the PRD. The main competitive advantage of CSs is their function as price breakers and capacity buffers. CSs usually lack proprietary assets and financial capital for investments in R&D. This makes them highly vulnerable to changes in markets, technology or to a financial crisis. It can be expected that operational flexibility is their main competitive advantage, while strategic or structural flexibility is less important. CMs are expected to have either strategic flexibility or, when upgrading and innovation activities are of greater importance, structural flexibility (ERNST 2004:95–96). As the subcontracting network of firms seems to have a great influence on the flexibility of firms in HK, the organisation of the network must be studied. This depends strongly on the institutional environment, which will be covered in the next Section 2.3.

2.2.3. Interim Conclusion and Hypotheses derived from Section 2.2

A *Which critical factors pushed the global spread of production locations, especially in the electronics industry? How does this affect the organisation of value chains? Which new challenges and risks do electronics firms in emerging economies face if they want to enter the global production network?*

The economic boom in the 1970s was the main driver of the development of global value chains. Increasing demand for electronics products, the advance in organisation and information technology, and the automatisation of production pushed the growth of the electronics industry. However, this development was accompanied by a high market volatility and short product life cycles. Leading Western firms had difficulties handling these factors. At the same time, the competence of suppliers in third-world economies increased and international trade barriers were removed. Leading Western firms started to outsource their production to large CMs which they shared with other LFs. Modularisation of production took place at steps in the production process where information could be easily codified. This production model of the electronics industry is referred to as a modular value chain. Firms in emerging economies had to compensate for LFs' weaknesses in dealing with demand changes, short production cycles and fast delivery times. Only firms providing a high flexibility were selected to supply LFs. Flexibility can be understood as the degree to which a firm has a variety of capabilities for reacting to demand changes and the speed at which those capabilities can be activated (VOLBERDA 1996:361). Integrating themselves into global value chains is identified as the main challenge for electronics firms in emerging economies. To prove propositions derived from the conceptual discussion, the following empirical research questions and hypotheses were formulated:

> **G** *How are electronics firms in HK and the PRD integrated into global value chains? How is the division of work spatially organised between HK and the PRD?*

Hypothesis A1 Electronics firms in the GPRD have increased their market share in a growing industry sector due to the outsourcing tendencies of leading Western firms.

Hypothesis A2 For electronics firms in the GPRD, the intensity of competition is high and increasing, markets are volatile and changes are virtually unpredictable.

Hypothesis A3 Large HK-based electronics firms are integrated into modular value chains as CMs.

Hypothesis A4 Production activities of electronics firms in the GPRD take place in the PRD. HK firms have kept their headquarters and higher-value activities in HK, but have shifted production activities and simple services to the PRD.

Hypothesis A5 The distance to LFs as customers is great, whereas the production networks between CMs and CSs work in spatial proximity.

Hypothesis A6 Electronics firms in the GPRD are characterised by quantitative and qualitative flexibility in order to respond to LF requirements.

2.3. INSTITUTIONS, TRANSITION AND INFORMALITY: A REGIONAL PERSPECTIVE

This section focuses on institutions, institutional transition and informal behaviour determined by the institutional environment. According to the model of LEWIN et al. (1999:537), this section deepens the discussion about the interplay of institutions on the regional level and the behaviour of firms. Besides the industry and market development, institutions also determine the behaviour of firms. In order to answer the research question **B** outlined at the beginning of this work, this section will:

- explain the interaction between the institutional environment and the informal behaviour of firms under special consideration of an institutional change in transitional economies

- discuss the term "informality" in the behaviour of firms and its effects on their flexibility

Section 2.3.1 will examine the formal and informal institutional framework on a regional level. Section 2.3.2 focuses on the transformation process in content and time in transitional economies. However, it will be argued that in transitional economies, informal institutions often determine the rules of behaviour. As informal institutions evoke informal practices in the behaviour of firms, Section 2.3.3 seeks to discuss "informality" and outlines how the term is interpreted in this work. Dimensions of informality which facilitate its understanding and measurement will be identified. Section 2.3.4 gives a brief overview of informal behaviour of firms in the applied research areas—recruitment, contractual arrangements and their enforcement in customer producer relations—and their expected effects on the flexibility of firms. The final Section 2.3.5 addresses the unresolved question of whether informality in a firm's interaction pattern can be taken as a substitute for a lack formal rules, which would suggest that it is a transitional phenomenon, or whether it is a complement to these rules and can be seen as a permanent phenomenon.

2.3.1. Playing the Game: Institutions and Institutional Change

Neoclassical economic theory was dismissive of institutions. This was one of its major weaknesses. It is now generally accepted that institutions are important for the economic development of regions and firms. According to NORTH (1990:3), "institutions are the rules of the game in a society or, more formally, are the humanly-devised constraints that shape human interaction." A method now prevalent is to study institutions using the new institutional economics (NIE). The NIE basically arose from changing assumptions about the neoclassical model. Proponents of the theory took a more realistic view of human behaviour than the traditional neoclassical approach (RICHTER and FURUBOTN 1996:35–38).

1. The neoclassical theory assumes *profit maximisation* to be the first goal of business. But the NIE identifies the importance of *institutionally expected utility*, which is based on more realistic human behaviour that can be directed by ideology, altruism and self-determined behaviour. Institutions are the key to understanding the complex rules of behaviour, because they help to explain the choice of business goals other than profit maximisation (for example gaining reputation, market power, technological upgrading, internationalisation) (WILLIAMSON 1987:45; NORTH 1990:26–60; RICHTER and FURUBOTN 1996:3).

2. The second assumption shifted from *complete rationality* to *bounded rationality*. This implies that players are not able to survey complete information and thereby comprehensively predict their business future. The environment is too complex and the cognitive competence to cope with all the information is limited (RICHTER and FURUBOTN 1996:3–5; NORTH 1990:17–26; SIMON 1961:xxiv; WILLIAMSON 1987:45–46).

3. The third shifting assumption concerns the behaviour of human beings. In the neoclassical model, human beings behave as "*homo oeconomicus*", but the NIE assumes that people do not automatically observe the given rules and try to act to their individual advantage. They behave *opportunistically*. Opportunism is a troublesome source of behavioural uncertainty in economic transactions (WILLIAMSON 1987:447–50; RICHTER and FURUBOTN 1996:4–5).

The core idea of the NIE is to emphasise the impact of institutions on the behaviour and decision-making processes of firms. This is related to the success of firms and therefore to regional economic development. Experts have tried to endogenise the variable *institutions* in economic models in order to measure its contribution (COASE 1992:714). This leads to the central research targets of the NIE—the description of institutions, the analysis of effectiveness of institutions and political recommendation for the establishment of rules with respect to efficiency.

WILLIAMSON (2000) developed four levels of analysis to describe the economics of institutions against a time-conscious background. He studied *informal institutions* at the first level, because they are deeply embedded in society (see Figure 2.5). Informal constraints are characterised by the customs, norms, traditions, culture and religion of a region. They are based on implicit understanding and are socially derived, which makes them inaccessible through written documents. However, it is widely recognised that it is easier to describe and to be precise about the formal rules that societies devise than about informal constraints. Informal institutions are path-dependent and have a long pervasive influence on the character of economic development. Institutional change on this level is gradual and happens over centuries or millennia. Many of them are linked with complementary formal institutions. The solid arrows connecting higher with lower levels indicate that the

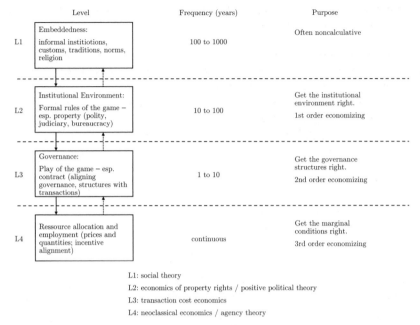

Level	Frequency (years)	Purpose
L1 Embeddedness: informal institiotions, customs, traditions, norms, religion	100 to 1000	Often noncalculative
L2 Institutional Environment: Formal rules of the game – esp. property (polity, judiciary, bureaucracy)	10 to 100	Get the institutional environment right. 1st order economizing
L3 Governance: Play of the game – esp. contract (aligning governance, structures with transactions)	1 to 10	Get the governance structures right. 2nd order economizing
L4 Ressource allocation and employment (prices and quantities; incentive alignment)	continuous	Get the marginal conditions right. 3rd order economizing

L1: social theory
L2: economics of property rights / positive political theory
L3: transaction cost economics
L4: neoclassical economics / agency theory

Figure 2.5.: Institutions and decision making in a regional aspect
Source: According to WILLIAMSON (2000:597)

higher level imposes constraints on the lower level. The reverse arrows are dashed and signal feedback. The second level can be seen as the *formal institutional environment*. Formal rules involve laws and regulations, for example property rights, constitutions (FURUBOTN and RICHTER 1991:3–6). Changes in formal institutions occur either over decades or centuries (for example the development of the EU) or when dramatic events result in sharp breaks with established procedures (civil war, occupations, breakdowns, financial crises). While informal norms only change gradually, formal rules can change overnight. According to NORTH (1990:3–5), the institutional environment is composed of formal rules, informal constraints and the enforcement characteristics of both. The formal constraints set the rules and regulations and the informal norms define the contour and the way in which the rules and regulations are specified and in which enforcement is carried out. Informal institutions provide a basis for self-enforcing contracts, as norms and conventions put pressure on people to stay with the concluded agreement. Formal institutions provide a third party *enforcement system* to enforce contracts (for example courts). On the third level, there is a *functional legal system for the conclusion and enforcement of contracts*, which influences the choice of governance modes (market, hybrid, firms). Transaction cost theory focuses mostly on explaining the existence of alternative modes of governance and the trade-offs among them against different institutional backings (see Section 2.4). COASE first mentioned the existence of transaction costs in his highly influential essay "The Nature of the Firm". He pointed out that a firm consists of a system of relationships. When the costs of establishing, maintaining and using the relationships (price system) become too high, the organisation of business activities under a central command may become advantageous (MENARD 2005:287; BICKENBACH et al. 1999:2). However, COASE's idea was extended by NORTH (1990, 1997) and WILLIAMSON (1971, 1975, 1985, 1996, 1997, 2005). They structured the concept of the NIE, because they saw that institutions were the true novelty in COASE's theory. The creation, organisation and use of institutions produces transaction costs. The contract theory examines the way in which contracts can be formed to regulate interests between agents most efficiently (see Section 2.5). The period over which rules regarding transaction costs are changed is one to ten years. To analyse governance modes in customer producer relations, the point of time when relations were established must be noted against the background of institutional change. Level two and three are typically analysed by the NIE. On the fourth level, *firm's behavioural adjustment* of governance modes is studied—resource allocation employment etc. Adjustments are made continuously (WILLIAMSON 2000:595–600; WILLIAMSON 1998:23–29).

It derives from WILLIAMSON (2000) that a change in informal institutions will be followed by a change in formal institutions. In reality, a change in formal institutions is not only driven by a change in informal institutions. Formal institutions can also be transferred from other economies. Many Western countries have developed a sophisticated system of rules and laws. Emerging and transitional economies tend to adopt entire law systems rather than develop their own system, which is costly and time-consuming. This formal institutional system is not adapted to existing informal institutions. Therefore, transferring the formal political and economic rules of successful Western market economies to emerging economies is not a sufficient condition for good economic performance. It is essential to change both the formal institutions and the belief systems for successful reforms (NORTH 1990). Otherwise, firms, led by individuals, will not support the new economic rules of the game, as their way of doing business is shaped by informal institutions and is not compatible with the new rules. There is empirical evidence to show that although formal rules are introduced, firms do not necessarily comply with them (MACAULAY 1963; DIMAGGIO and LOUCH 1998; DE MESQUITA and STEPHENSON 2006). It is not the existence of formal institutions, but rather the costs, the efficiency and the willingness to use them which makes them attractive for firms. This explains why the economic system in some nations still does not work effectively, even though formal institutions are well developed. Formal rules only take effect if they are interwoven with the existing informal constraints. GRANOVETTER (1973) argued that such informal constraints still form the basis of economic relationships even in modern economies.

In China, the legal system has been transformed, but the way in which firms do business has not yet adapted to the new rules of the game. Therefore, it seems worth studying how firms organise their business in an institutional environment in which formal and informal rules are not perfectly interwoven. With the increasing complexity of societies, formal institutions complement, revise or replace the effectiveness of informal constraints.

2.3.2. Regions in Transition

When regions transform from a centrally planned economy to a market economy, as China is doing, they design a set of formal rules which need to be introduced to combat problems caused by the former leadership. A transition from planned to market economies implies the transformation from fixed prices and production plans to a supply and demand oriented production. For this reason, prices could not be taken as a stimulation measure of economic competition. A lack of product quality was evident. The collectively-owned capital goods did not stimulate firms to maintain or upgrade their capital stock. Both deficits account for the inefficient organistion of communist economies (LAVIGNE 1995). Moreover, firms did not need to compete on the market because of the foreign trade monopoly held by the state. The reform of the central planning system has aimed to establish political stability and pluralism through a competitive legal and administrative system, creating basic financial institutions of capitalism and implementing a programme of real economic reforms with the evolution of a full market economy. There is no panacea or universal solution for a transformation process, but the change of certain elements and actions has been commonly agreed on (HUNTER and RYAN 2008; REVILLA DIEZ 1995).

1. **Reform of the legal and administrative system**

   - Establishment of a *legal economic security system*—trade regulations (new trade laws, membership of OECD, WTO), property rights, bankruptcy law, competition law

   - Revision of the *tax code*—enterprise tax, income tax

   - Reform of the *administrative system*—degradation of bureaucracy, development of public authorities

2. **Reform of the financial system**

   - Reform of *price formation*—liberalising prices from state control, stabilising inflation, new system of public finance

- Reform of the *banking system*—opening up for foreign banks, establish-ment of a functional banking system

- Reform of *foreign capital trade*—full convertibility of the local cur-rency, free foreign exchange and capital transfer

3. **Establishment of new market institutions**

- Reform of the *enterprise sector*—open market for the establishment of private enterprises, privatisation of state-owned enterprises, open market for foreign direct investments

- Development of a *stock market*

In implementing these changes, two reform strategies may be identified and distin-guished: slow or gradual transformation (for example China) and rapid or shock tactics (for example Poland, Estonia). Focussing on the customer producer rela-tions, not all transitional issues mentioned affect the organisation of trade. The most important is the establishment of a functional *legal economic security system*, especially trade and property rights. Firstly, this leads to a higher attraction and in-crease of foreign trade (more customer producer relations) and secondly, firms can rely on contracts and their enforcement. If those rules are not functioning, firms avoid the market or rely on informal means to overcome its weaknesses. Equally important is the reform of the enterprise sector to allow the establishment of *pri-vate foreign enterprises*. Moreover, the reform of the *foreign capital trade* system is essential, because firms want to exchange not only goods, but also capital.

In economic and political awareness, indicators for institutional change are usu-ally the development of formal market institutions. Less attention is paid to the fitness of formal and informal institutions. In transitional regions especially, dif-ficulties arise from the interplay of formal and informal institutions. Therefore, YEUNG and LIN (2003) invite a context-specific theorising of the economic ge-ographies of Asia, bearing in mind those difficulties. It is still of importance how far the institutional framework is developed according to the points indicated above, but attention should also be drawn to how well the new reforms are applied in actual business practices. Informal institutions might still shape the way of doing business, as they co-exists with formal institutions. The focus should be set not only on the introduction of formal institutions, but also on their enforcement.

Outlook: Formal and Informal Institutions in HK and the PRD

The Chinese reform policy began in 1979 with the introduction of the first market-oriented regulations. By now, the Chinese government has managed to revise most of their laws concerning the economy. However, businessmen in China still per-sist with their former business practices based on more informal exchanges. This is due to the persistence of informal constraints despite the introduction of formal rules. Section 4.1.1 in the empirical part will focus on the formal institutional change and the complementary Section 4.1.2 will outline the informal institutional environment shaping the way of doing business in China. The empirical part will reveal why Western rules introduced in emerging economies differ in performance characteristics from the original country.

In contrast to China, the HK territory was shaped by the British from the 19th century onwards. Western rules and laws were introduced at an early stage. HK firms are now very used to dealing with Western institutions. Simultaneously, HK entrepreneurs know about the Chinese norms and values (informal institutions) shaping their way of doing business. As firms in HK act under a different institutional setting from firms in the PRD, they are expected to organise their business differently.

Firms in the GPRD are characterised by cross border business organisation, whereas the border between HK and the PRD separates two distinct institutional environments. HK firms are familiar with both business systems. They are expected to take the role of intermediaries. This requires knowledge of formal and informal institutions in both regions. HK firms are assumed to benefit from this implicit knowledge while applying a sophisticated mix of formal and informal modes in interactions with customers and producers. Section 4.1 will provide an overview of the institutional settings in both HK and the PRD.

2.3.3. Benefiting from Informality in Interactions

In the literature, the term informality is often vaguely defined and a plurality of interpretions exists. Initially, informality referred to illegality. Later, the phenomenon changed and informality appeared in conjunction with the informal economy. It referred to unregulated labour-intensive activities, self-employed entrepreneurs, micro and small enterprises, activities taking place outside state regulations, unregistered activities, etc. In recent literature, informality has been related to informal arrangements and behaviour within formal firms or organisations (LI 2007; SINDZINGRE 2006).

A distinction must be made between the perspective of institutions creating formal and informal rules for the game on the one hand, and the perspective of formal and informal behaviour of firms guided by the rules of the game on the other hand. This work emphasises the distinction between formal and informal behaviour in terms of customer producer relations. The perspective of formal and informal institutions only serves as a framework to explain the behaviour of firms. It is not assumed that formal behaviour follows formal rules or the other way around. Informal rules can also determine formal behaviour, for example when it is commonly agreed that contracts need to be written, although this is not required by the lawmakers. Conversely, formal rules can determine informal behaviour, for example when

Table 2.1.: Formal informal dimensions in business interactions

Dimensions	Formal	Informal
Personalisation	impersonal	personal
Form of agreements	written	verbal
Preciseness of agreements	precise	open-ended
Enforcement of agreements	tight	loose
Power	legal (vertical)	social (horizontal)

Source: Own composition according to LI (2007:229–231)

parties argue, it is mandatory in some countries to work on an informal agreement before they are allowed to go to court. LI (2007) discussed the contrasting terms of formality and informality of firm's behaviour. He distinguished between formal and informal attributes of interaction, whereby these mark the extreme points of a continuum. According to LI (2007:229), informality in interactions on the micro-level

> " refers to the nature of social ties and events as implicitly assumed, endogenously embraced and flexibly enforced by horizontal [...] pressures in a particularistic personalized process [...], while formality as explicitly prescribed, exogenously imposed and rigidly enforced by vertical authority powers in a [...] depersonalized process."

To study formal and informal behaviour, LI proposed dimensions of informality as a framework to study and measure empirically formal and informal practices in business interactions. He recommended taking into account the degree of *personalisation* in business interactions, the *form of agreements* concluded, the *precision of arrangements*, the *enforcement mechanisms* of contracts and the *power* of institutions used (see Table 2.1).

LI divided each dimension into a variety of actions ranging from formal to informal. Business contacts can be highly *impersonal* (formal) or very *personal* (informal). Personal processes are based on face-to-face contacts, trust between partners and private involvement. This may also determine the form of agreements. Contracts can be highly formalised in a *written* way or informally agreed on in a *verbal* way, when partners trust each other's word. Agreements can be taken as strongly formal when they are completely in writing, whereas they are less formal when the key points are fixed in writing but complemented by further points verbally agreed on. Agreements between parties which involve no writing are strongly informal. Agreements which consist of rough written points, but verbally agreed details and specification are less informal. Not only the form of agreements, but also the precision of agreements can range from very *precise* (formal) to *open-ended* (informal). Business culture can rely on open-ended agreements with permanent renegotiations instead of specifying details in advance. Formal and informal agreements rely on different enforcement mechanisms, which can either be very *tight* on one extreme and very *loose* on the other. In the case of conflicts, the power used to finally enforce agreements can be *legal* via courts or *social* power relying on shared norms within a culture (LI 2007; SINDZINGRE 2006). A firm can be characterised as acting on informal constraints in terms of customer producer relations when dimensions tend to be specified by informal practices. In contrast, a firm is characterised as acting on formal constraints in terms of customer producer relations when dimensions tend to have a formal shape. In Section 2.5, the dimensions will be discussed in more detail with respect to their appearance in the recruitment of customers and producers, contractual arrangements and enforcement of contracts.

2.3.4. Informality reaching Flexibility

It can be expected that firms which apply informal practices in interactions with their customers and producers gain business advantages resulting in more flexible means of reacting to market and industry changes. According to Section 2.2.2, flexibility combines the variety of options for reacting and the speed with which they can be applied. In this section, the different dimensions of informality—personalisation, form and precision of agreements, enforcement and power—and their contribution to flexibility are taken into account. A special emphasis is put on personalisation. It can be expected that personal relationships are the basis for the informal shape of the other dimensions of informality. Personal trust, reliability, reputation and mutual understanding ensure and safeguard business in social ways. Social safeguard mechanisms are a precondition for a verbal and less precise agreement or a loose enforcement of contracts (POWELL 1990; DIMAGGIO and LOUCH 1998). Whereas it can be expected that social ties can reduce the time required to conclude contracts due to the fact that personal trust smooths businesses relations, one has to consider that it takes a lot of time and effort to build social ties. Besides, social ties can also lead to lock-in effects, for example when firms are forced to select customers and producers via social ties instead of choosing the highest quality partner on a market basis. In other cases, firms cannot enforce contracts without considering the fact that this could burden the relationship. Lock-ins can negatively affect the competitiveness of firms (LORENZ 1988).

The interplay and dependency of the five dimensions of informality will be studied according to the three research areas—recruitment, contractual arrangements and enforce mechanisms of contracts in customer producer relations. The recruitment process will be studied to provide an idea of the degree of personalisation of the relationships. This is essential to understand how firms manage the second research area, contractual arrangements, regarding the form and precision of agree-

Table 2.2.: Informality reaching Flexibility

Research Areas	Dimension of Informality	Flexibility	
		Variety	**Speed**
Recruit-ment	Personalisation	The formal options to get access to partners can be extended by additional personal ways to contact partners (e. g. bidding competition vs. personal networks).	The formal process to recruit customers and producers can take more time (deadlines, bureaucracy), whereas the personal network is always ready when demand is there, but it needs much more time to build. Time-consuming investments in the beginning lead to quicker business operations when demanded.
Contractual arrange-ments	Form and preciseness of agreements	When relations to customers and producers are on a personal base, trust is developed which implies informal safeguards for contracts. Contracts which are verbally concluded and less precise can easier be renegotiated, refined, and adopted in a variety of ways to changing economic circumstances.	The process of contract conclusion and renegotiation is quicker when details do not need to be worked out in writing.
Enforce-ment Mecha-nisms	• Enforcement of agreements • Power system	The enforcement of contracts by social power extends the variety of options firms can use to settle disputes (e. g. negotiations vs. litigation).	Using social power and loose enforcement mechanisms is quicker than a formal litigation process, but social power also involves the risk not to find any solution, because no clear judgment can be made.

Source: Own composition

ments. Likewise, the applied enforcement mechanisms and the power system in terms of conflict solutions, e. g. the third research area, depend on the degree of personalisation of relationships. In the following Table 2.2, the three research areas and the corresponding dimensions of informality are summarised. Furthermore, it is illustrated how an informal shape of those dimensions contributes to flexibility in terms of variety and speed under consideration of the negative and positive effects of informality.

2.3.5. Interplay of Formal and Informal Institutions in Transitional Economies

Despite the common recognition that informal institutions influence business operations, there is a wide range of academic literature about their actual importance. Opinion ranges from them having only minor effects to them being very influential. HART (2001) contends that although norms are undoubtedly very important, it is still difficult to incorporate them into the theory of firms. In his opinion, norms have not added a great deal to understanding the determinants of firm boundaries. Authors such as WILLIAMSON (1991) acknowledge the role of informal institutions, but still treat them as exogenous sources that smooth formal institutions. Experts with a background in sociology stress informal mechanisms as being the central way of governing exchanges both inside and outside the firms (GRANOVETTER 1973; POWELL 1990). Although the different perspectives exist, in almost all studies conducted, formal institutions have been analysed and evaluated quite independently from informal institutions and the other way around.

A detailed view shows that those differences complement each other. HART (2001) might be right in his opinion that the influence of informal institutions on governance modes is low. The decision to choose a certain governance mode has serious consequences for firms, because once a mode is selected, it cannot be changed or revised easily. Therefore, firms tend to take into account formal rules rather than informal constraints for their decision. In contrast, when a governance mode is selected, informal practices help smooth relations. Along with formal mechanisms, they provide one way to practice business. They influence a firm's interaction with customers and producers in terms of recruitment, contracting and enforcement. Formal institutions still dominate crucial business decisions, but the influence of informal institutions is important for running the business smoothly. From an economic point of view, the concentration on informal institutions, as is the preference in the field of sociology, seems inappropriate for the analysis of customer producer relations.

There are some experts who approve of the view that formal and informal institutions act as substitutes for one another. CAI and YANG (2008) argue that well developed informal institutions can take over the functionality of formal institutions. Instead of safeguarding business by law, it can also be safeguarded by social norms and trust. In countries where the formal laws and regulations do not fully safeguard transactions, informal institutions can substitute them. Some experts take an even more extreme position, arguing that formal constraints are unnecessary because informal relationships based on trust and social norms can support cooperation without the costs and complexity associated with formal arrangements (UZZI 1996; POWELL 1990). Informal institutions reduce the benefit of using formal institutions. Others argue even more strongly and claim that formal modes are not only unnecessary when informal practices are applied, but also affect them in a negative way. Legal security and formal contracting, for instance, can erode the interpersonal relationship.

An alternative argument is that informal institutions can also complement formal mechanisms. This view is more common in the academic literature and could be proven empirically. In countries where the functionality of the formal institutional environment is ensured, firms combine transactional safeguards of formal and informal institutions to achieve the best results. As contracts are always incomplete, parts which cannot be formalised will be settled on an informal basis. If one part is formalised, it facilitates the enforcement of the non-contractable elements (ZENGER et al. 2002). Another argument is that formal institutions can set the stage for the development of trust for long-term interactions. Formal detailed contracts shift from specified deliverable outcomes to framework agreements for bilateral adjustment. This provides the basis for intensive cooperation. Additionally, informal elements may increase the performance of formal institutions. They provide ideas for the refinement, specification and optimisation of formal contracts in the next phase. ZENGER et al. (2002) conclude that "formal and informal institutions are not mere alternative ways to govern exchanges. In most cases they are employed simultaneously and interact in complex ways."

When institutions are in transition, it can be expected that in the absence of a functional formal institutional framework, informal institutions substitute formal institutions. But as soon as the formal framework develops, informal institutions change to complements. A transaction will always be governed by informal and formal constraints. If informality prevails in institutional transition, then informal interactions in business are not transitional, but permanent. This leads to the assumption that informality in business interactions overcomes the institutional constraints. It can be expected that crucial decisions regarding customer producer relations are oriented towards formal institutions. But it can also be assumed that informal business interactions retain their importance even after the transformation is completed, because they provide more efficient outcomes. To conclude, HK and PRD-based firms are expected to decide on different governance modes (as it is an essential decision), because the formal institutional environment differs between the regions (see Section 2.4). In contrast, it is expected that HK as well as PRD-based firms apply informal modes of interactions in their daily business (see Section 2.5).

2.3.6. Interim Conclusion and Hypotheses derived from Section 2.3

B *How does the institutional environment in a region determine the behaviour of firms under special consideration of institutional change in transitional economies?*

Formal and informal institutions govern the behaviour of firms in organising their customer producer relations. When countries undergo an institutional change, for example from a centrally planned economy to a market economy, as China is doing, the formal institutional environment changes through the introduction of new laws and regulations. According to WILLIAMSON (2000:597), the change to informal institutions follows the change to formal institutions with a time delay. Therefore, businessmen in transitional economies do not adapt immediately to new laws and regulations, but stay with their traditional, more informal business culture for a certain time, as the formal institutional environment is still incomplete or its functionality diminished. Informality is therefore not associated with illegality or the informal sector, but can be understood as informal interactions within formal organisations.

Firms' patterns of organising their customer producer relations can range from very informal to very formal, referring to the dimensions of informality: *personalisation, form* and *precision of agreements, enforcement mechanisms* and the *power* used to enforce agreements (LI 2007:229–231). In this case, informal behaviour completes formal behaviour. What is expected, but remains to be studied, is whether informality in interactions vanishes in complete formal institutional settings, or whether it remains important to a certain degree because it overcomes the paradigm of formal and informal institutions. Hence, HK firms would still apply informal modes to organise their customer producer relations. To prove propositions derived from the conceptual discussion, the following empirical research question and hypotheses are formulated:

H *How well is the formal institutional environment in HK and the PRD developed? What informal institutions guide the way of doing business in HK and the PRD?*

Hypothesis B1 China's formal institutional environment still has weaknesses concerning the safeguards for trade between firms.

Hypothesis B2 If the formal institutional framework changes from a planned economy to a market economy, informal institutions persist for a certain time. The business culture of HK entrepreneurs has, in general, adapted to the formal market mechanisms, whereas the Chinese managers still use informal practices more intensively than HK firms.

Hypothesis B3 Despite HK's complete establishment of formal institutions, informal elements in interaction with customers and producers remain important to a certain degree and complement formal practices.

Hypothesis B4 The use of informal practices in customer producer relations leads to greater firm flexibility.

2.4. GOVERNANCE MODES IN CUSTOMER PRODUCER RELATIONS: A FIRM'S PERSPECTIVE

In accordance with LEWIN et al. (1999), it has been discussed how the market and industry conditions on a global level (see Section 2.2) and the institutional environment on a national or regional level (see Section 2.3) influence firms' behaviour. This section is aproached from a firm's perspective and emphasises the adaptation of firms to a changing industrial and institutional environment. The focus is placed on the governance modes selected by firms to organise their customer and producer relations. To answer research question **C**, this section will give explanations as to:

- why transaction costs define the boundaries of a firm

- how transactional specifications derived from the market and industry conditions influence the choice of governance modes

- how the institutional environment determines the choice of governance modes

- what governance modes firms can be expected to apply in the GPRD

In Section 2.4.1, the concept of transaction costs and their importance for the governance modes of customer producer relations will be explained. Firms tend to choose the governance modes which result in a minimum of transaction costs. Section 2.4.2 refers to the different governance modes: market, hybrids and hierarchies. It analyses their strengths and weaknesses comparatively. As the selection of governance modes depends on the transaction costs involved, it will be taken into account which determinants influence those costs. In Section 2.4.3, transactional specifications will be examined according to their impact on governance modes. Section 2.4.4 explains the interplay of the development status of the institutional environment and the governance modes selected by firms. The last Section 2.5.1 gives an account of the special situation in HK and the PRD. Finally, the hypotheses derived from this section are presented.

2.4.1. Idea and Concept of Transaction Cost Economics

As outlined in Section 2.3.1, COASE (1937) first introduced the idea of transaction costs to the academic community. Traditional economic theory suggested that because the market is "efficient", it should always be cheaper to subcontract than to produce in-house. But instead of using markets, companies tend to be organised as hierarchies, using a chain of command and control rather than negotiation, markets and explicit contracts. COASE (1937) explained that the costs of using the market—he calls them transaction costs—were too high. Within the firm, the entrepreneur may be able to reduce these transaction costs by coordinating these activities himself. WILLIAMSON (1987:19) sees transaction costs as "the economic equivalent of friction in physical systems". When firms organise their customer and producer relations, initial transaction costs occur when arranging and organising the transaction in advance (*ex-ante transaction cost*), and subsequent transaction costs arise for the implementation of transactions afterwards (*ex-post transaction cost*) (WILLIAMSON 1987:20–22). *Ex-ante transaction costs* are incurred in the following areas:

- **Information**: Costs occur when a firm is searching for customers and producers. Firms have to find information about experience, ability, reliability, location and contact persons of prospective partners. This includes costs for visiting potential partners, advertising, exhibitions etc. Firms can reduce their transaction costs using former partners who have already proven their reliability. When firms look for new customers and producers, business networks help to gain information and ensure against opportunistic behaviour of potential partners. Recommendations within business networks, for example, facilitate the access to potential partners and information (BLUMBERG 1998:18–19; RICHTER and FURUBOTN 1996:51–52).

- **Negotiation, drafting and conclusion of agreements**: This includes costs for preparation of information, legal advice, contract draft etc. The more complex the topic and the more money involved, the more difficult and time-consuming the negotiations, drafting and conclusion of contracts are. The precise description of contractual details in writing (formal) involves high costs, but makes enforcement costs calculable. In contrast, less precise and verbally agreed contracts (informal) involve lower ex ante costs, but make potential enforcement costs incalculable (BLUMBERG 1998:18–19; RICHTER and FURUBOTN 1996:52; LYONS 1994, WILLIAMSON 1987:20).

The *ex-post transaction costs* of working with customers and producers include the costs of running and control transactions, the cost of renegotiations, and the cost of the enforcement of contracts:

- **Running and control**: Costs arise from the necessity to control quality and quantity of products as well as the delivery time. Opportunistic behaviour has to be reduced to a certain limit. Trust, reputation and reciprocity—typical traits of business networks—decrease the costs of controlling transactions accurately.

- **Renegotiations**: Costs occur when transactional details have to be changed, renegotiated and refined. Renegotiations are either caused by unexpected changes in demand, or can be knowingly provoked by former incomplete contracts (KLEIN 2005:436).

- **Enforcement**: Dispute resolution and the enforcement of contracts involve costs, as can be seen in litigation processes. But alternatively, more informal forms, such as negotiation and mediation, can keep these costs to a minimum (GÖBEL 2002:64; RICHTER and FURUBOTN 1996:52–53).

WALLIS and NORTH (1986:121) tried to measure the transaction costs as a share of the GDP in the United States. According to their estimation, transaction costs rose from 16.1% in 1870 to 54.7% in 1970. Almost all the empirical literature avoids any attempts to measure transaction costs directly. Instead, they use a reduced model in which transaction costs are assumed to be minimised by firms. WILLIAMSON (1987) argues that transaction costs vary depending on the institutional environment, because the use of institutions involves costs. As firms try to minimise transaction costs, they adopt firm organisation to the institutional environment. They look for the most efficient way to organise their business with customers and producers. Basically, the academic literature distinguishes between three governance modes—market, hierarchy and hybrids—organising the relationships between customers and producers. In reality, many more mixed forms can be observed (WILLIAMSON 1979:235). As firms tend to minimise their costs, they select the optimal governance mode. For example, when market institutions are not well developed, their use is very costly for firms. However, firms tend to avoid market relations and organise their customer producer relations in hybrid modes or hierarchies. The next Section 2.4.2 addresses the traits associated with the different governance modes (MENARD 2005:293; SHELANSKI and KLEIN 1995:285).

2.4.2. Traits of Markets, Hierarchies and Hybrids

Markets

From the economic point of view, markets should be the most efficient governance mode to organise the production process and herewith relations to customers and producers. Markets are coordinated by prices. Therefore, they remain highly impersonal in comparison to hybrid forms or hierarchies (MENARD 2005:303–305). The existence of a large number of potential trading partners provides an extensive choice on the one hand, but on the other hand also increases the cost of finding the right trading partner (BICKENBACH et al. 1999:6). Customers and producers can be randomly paired and exchanged (see Table 2.3). Therefore, specific investments are rare on markets. Market transactions provide players with considerable autonomy and flexibility to exploit profit opportunities by adjusting their behaviour to unfolding events. But in the case of major events which require the adaptation of the entire value chain, market-based chains have difficulties adapting because of their anonymous organisation. Contracts constitute an important arrangement for organising market transactions, since firms do not have much else for the parties to rely upon. Different governance modes use distinct forms of contracts. The classical contracts characterised by legal rules, formal documents and self-liquidating transactions are typical for market transactions (MENARD 2005:304–305). Therefore, effective third-party enforcement mechanisms (public ordering) are required, which are usually associated with a market-oriented institutional setting (MENARD 2005:304–305). Markets rely heavily on a formal setting according to the definition of informality. The relationships are usually organised in impersonal ways, and contracts are therefore mostly written in detail and enforcement is organised by courts.

Hierarchy

As mentioned above, at the other end of the spectrum lies the hierarchical exchange. It characterises transactions that take place under the unified ownership and control of one firm (vertical integration). When investments are very specific, vertical integration is often the best way to protect investments against opportunistic behaviour. The biggest disadvantage of hierarchies is their low-powered economic stimulation, which reduces their ability to adapt quickly to unpredictable market changes. Moreover, the limited external exchange of information and knowledge can lead to lock-in effects (see Table 2.3). In contrast, hierarchies provide relatively efficient mechanisms for responding to major market changes where coordinated adaptation of several units of a value chain is necessary (adaptation of entire units). As firms do not need proper contracts for internal firm organisation, managers can easily react to orders and have the right to reallocate tasks (MENARD 2005:291; KLEIN 2005:436). It can be expected that HK firms opt more often for a hierarchical organisation when working in China, because it seems to be suitable even if environments are no longer so uncertain. WILLIAMSON (1998:50) refers to a third mode of organisation—the T-mode, where T indicates a temporary or transitional situation. This form can be observed in developing markets where technology and rivalry are undergoing rapid changes. According to WILLIAMSON (1998:50) "Joint ventures [...] should sometimes be thought of as T-modes of organization that permit the parties to remain players in a fast-moving environment." Parties can pool resources to meet market demand for price and quality. Unsuccessful joint ventures (JVs) will transform to other governance modes later on, while successful JVs will remain in operation. In this work, T-modes are counted among hybrid forms of organisation (see next paragraph). It can be expected that during economic transition in China, firms tend to rely more heavily on those intermediate forms to keep their flexibility.

Hybrids

In between markets and hierarchies, the extreme modes of governance, there is a broad range of hybrid governance structures. Examples of hybrid modes include

Table 2.3.: Attributes of governance modes

	Market	Non-equity Cooperation	Equity Cooperation	Hierarchy
Theoretical nature	arm's length contracts	commitment to cooperate	obligation to cooperate	vertical integration
Character of relations	contracting arrangement on buying and selling	framework agreement	registered exchange of equity	establishment of subsidiary by parent company under unified ownership
Intimacy	impersonal	personal	personal	internal relations
Asset specificity	low	medium low	medium high	high
Frequency of transactions	low	medium low	medium high	high
Investment	not specific	moderately specific	specific	very specific
Change of customers and producers	very easy	easy	difficult	nearly impossible
Administrative control	nil	little	some	much
Quick response to external changes	very high	high	low	very low
Adaption of entire units	very low	low	high	very high
Enforcement mechanisms	legalistic (public ordering)	mixture of public and private ordering	mixture of public and private ordering	private ordering

Source: own composition according to WILLIAMSON (2005:49); PIES (2001:15) and WANG and NICHOLAS (2007:131–137)

the exchange of shares with trading partners, a joint ownership arrangement, the issuing of a licence to another firm, long-term contracting (framework agreements), franchising, strategic alliances etc. (SHELANSKI and KLEIN 1995:341; KLEIN 2005:438; MENARD 2004:347–350). Hybrids develop when markets shape up as being unable to adequately allocate the relevant resources and capabilities in situations where vertical integration would reduce flexibility, create irreversibility and weaken market stimulation. In choosing a specific form of hybrid, contracting parties attempt to retain the respective advantages and avoid the respective disadvantages of markets and hierarchies for transactions. According to MENARD (2004), three regularities characterise hybrids: pooling of resources, contracting modes and competing. The capitalisation on pooled resources and capabilities requires inter-firm coordination and cooperation. This involves the risk of opportunistic behaviour. Therefore, the identity of partners is important. Hybrids involve joint planning and an exchange of codified and tacit knowledge, competencies and technologies. It can be expected that firms which produce complex products for the high-end market rely more often on hybrid firms than on hierarchies or markets (WANG and NICHOLAS 2007:131–137). They want to avoid opportunistic behaviour—a risk of markets. Additionally, they want to facilitate external knowledge transfer, which is difficult in hierarchies. Pooling resources does not make sense without some continuity in their relationship. This leads to different contracting modes—the second regularity. Classical contracts tend only to provide a relatively simple and uniform framework. They are less suitable for serving hybrids in long-term relationships. This assumes that hybrids rely much more on personal issues, because otherwise the risk of opportunistic behaviour is too high. Without a certain degree of informal constraints—trust, reputation, reciprocity—hybrids do not work efficiently (BICK-ENBACH et al. 1999:7–8; SHELANSKI and KLEIN 1995:345; MENARD 2005:296). Therefore, hybrid modes are taken as more informal modes of governance in this work. Section 2.5.1 provides more insights into how informality in hybrids is applied in China (guanxi). A third determinant is the role of competition and cooperation among partners. Although they cooperate on some issues, parties also compete against each other. The risk results from difficulties in changing trading partners and unforeseeable revisions regarding the cooperation. Moreover, despite the advantage of limited opportunistic behaviour in long-term cooperation, firms have to bear in mind that they sometimes miss out on good deals from other firms (MCMILLAN 1995:213). The traits of hybrids seem to be characterised by intimacy, privacy and interdependency of partners. Different forms provide firms with different opportunities for doing business. The involvement of equity in hybrid modes seems to distinguish two fundamental forms of hybrids:

- **Equity cooperation:** Relations involving capital exchange between partners (for example JVs) are referred to as equity cooperation (see Table 2.3). Equity cooperation is more closely related to hierarchies than to markets. The common ownership provides an incentive for non-opportunistic collaboration and facilitates the transfer of resources between partners. The complex procedure of registering an equity exchange involves a long-term perspective (WANG and NICHOLAS 2007:131–137).

- **Non-equity cooperation:** Hybrids which are organised without any equity relationships are referred to as non-equity cooperation. Their affinity for cooperation can be indicated by framework agreements, which can be loosely concluded. Non-equity cooperation is long-term oriented to reduce uncertainty and support investments (GOFFIN et al. 2006:191). This is different to buying-selling agreements on markets. The non-equity cooperation is more closely related to markets than to hierarchies. WANG and NICHOLAS (2007) emphasised the role of equity and non-equity cooperation, especially against the background of research in the GPRD. Incorporation in non-equity cooperation means more openness and greater flexibility, allowing terms and conditions to be continuously negotiated between partners, but without being unified under common ownership. Such organisational structure allows HK firms, for example, to exercise control over Chinese producer firms without incurring equity investments. Non-equity cooperation is characterised by a strong commitment of parties to exchanging information, sharing control and making common decisions. But as no capital exchange has to be registered, transaction costs of entering and exiting the partnership are attenuated. Non-equity relationships can be established quickly in order to take advantage of short-term business opportunities and dissolved when they have completed their assigned task (WANG and NICHOLAS 2007:131–137).

2.4.3. Transactional Specification and Governance Modes

Transaction cost economics assume that there are rational economic reasons for organising transactions in one way or another. The literature distinguishes between specifications concerning the transaction and specifications concerning the institutional environment as critical factors for determining the governance mode to organise relations to customers and producers (see Figure 2.6).

Assuming that the institutional setting remains unchanged, transaction specification should be taken into account first. The attributes discussed in the literature which have a significant influence on the selection of governance modes are the *degree of relationship-specific assets*, *market and behavioural uncertainty* and the *frequency* with which the transaction occurs (SHELANSKI and KLEIN 1995:338; MENARD 2005:285). In order to compare alternative governance modes, the analysis focuses on three attributes (independent variables) which determine the mode of governance with the minimal transaction cost (dependent variable). In the following section, the three attributes will be examined in detail. **Asset specificity** is the most influential attribute for transactions. According to WILLIAMSON (1987:55) "asset specificity refers to durable investments that are undertaken in support of particular transactions". In fact, it contains the value of investments which would be lost in any alternative use. Highly specific assets create mutual dependence. The academic literature mainly distinguishes between physical asset specificity and human asset

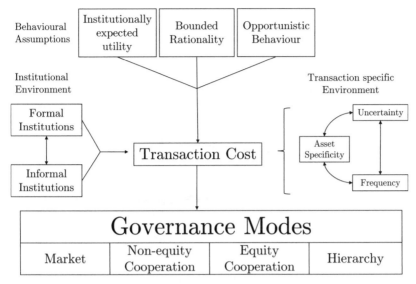

Figure 2.6.: Determinants of transaction costs and governance modes

Source: Own figure

specificity (SHELANSKI and KLEIN 1995:339). *Physical asset specificity* refers
to special purpose equipment for production. *Human asset specificity* describes
the worker-specific knowledge, transaction-specific knowledge or human capital.
It can be achieved through specialised training or through learning-by-doing. Human asset specificity can be measured by technical specifications such as product
complexity, the degree of innovativeness of firms or the qualification of workers
(WILLIAMSON 1979:242; SHELANSKI and KLEIN 1995; KLEIN 2005:436–439;
SAUSSIER 2000:385). When asset specificity is high, firms tend to opt for a hierarchical relationship to strongly control and govern production. If asset specificity is
low, firms tend to use market relationships.

Asset specificity gains importance in conjunction with the occurrence of bounded
rationality and opportunism, because it leads to **uncertainty** about future events. In
the transaction cost theory, there is a distinction between market and behavioural
uncertainy. *Market uncertainty* derives from volatile and unpredictable market conditions (REICHHART and HOLWEG 2007:1153). The more predictable the changing
demand for firms and the lower the competition intensity is, the better firms can cope
with market uncertainty (KRAJEWSKI et al. 2005:454; KLEIN 2005:436–439). In
addition to market uncertainty, *behavioural uncertainty* has been taken into account
(WILLIAMSON 1987:58). The probability of behaving opportunistically cannot be
measured in advance because of the bounded rationality. One can attempt to assess behavioural uncertainty using industry experience working with partners, the

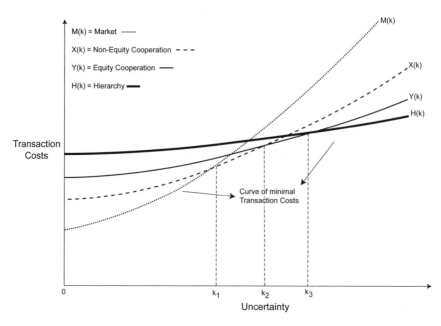

Figure 2.7.: Trade-off between markets, non-equity cooperation, equity cooperation
        and hierarchies

Source: Own draft based on PIES (2001:17), translated

method of choosing partners or the precision of contractual agreements. In long-
term dense cooperation, the ability to anticipate the behaviour of the other party
increases. Better knowledge and communication within customer and producer relations leads to reduced uncertainties. Figure 2.7 demonstrates the influence of
market and behavioural uncertainty in transactions on governance modes. Uncertainty is measured by k, whereas $k_0 = 0$ indicates that no uncertainty is involved and
k_3 indicates the highest uncertainty. It follows that $k_0 < k_1 < k_2 < k_3$. A pure *market relationship* $M(k)$ is appropriate if $k_0 \leq k < k_1$. When markets and behaviour
are not uncertain, transactions do not require special safeguards. The advantage of
market stimulation and quick adaptation can be realised (see Table 2.3). It follows
that when only a few safeguards are needed, markets provide an efficient mode to
manage customer producer relations. k_1 indicates the trade-off point between markets $M(k)$ and *non-equity cooperation* $X(k)$. When firms face greater uncertainty
regarding the reliability of customers and producers ($k_1 \leq k < k_2$), they can either
opt for more complex contracts and safeguards, which involve higher transaction
costs, or they can transform their governance mode to partnerships in non-equity
cooperation, which provides them with the minimal transaction cost for higher un-

certainty. Although there is a widespread agreement that the discrete transaction paradigm "sharp in by clear agreement, sharp out by clear performance" has served both law and economics well, there is increasing awareness that many complex contracts are not of this well-defined kind (WILLIAMSON 1979:235–248; BICK-ENBACH et al. 1999:6; MENARD 2005:305). In non-equity cooperation, specific investments can be protected by personal trust and commitment. This results in the conclusion of framework agreements as a sign of long-term cooperation and trust in the fact that remaining points can be settled harmoniously at a later date. When the degree of uncertainty increases further ($k_2 \leq k < k_3$), *equity-based cooperation $Y(k)$* provides the lowest transaction costs for firms. Equity cooperation between firms and their customers or producers provides even more safeguards than non-equity cooperation. The unified ownership with a long-term perspective partner commits firms to invest in and support the joint success. The expectation of better safeguards compensates the burden of a complicated and expensive procedure to established a jointly owned firm. If the market and behavioural uncertainty turns out to be very high ($k \geq k_3$), *hierarchies $H(k)$* are the most cost efficient mode of organisation. Hierarchies offer greater protection for specific investments and provide relatively efficient mechanisms for responding to change where coordinated adaptation is necessary. A growing degree of uncertainty in terms of transactions makes the protection against opportunistic behaviour more expensive in hybrids and markets. These costs for protection against opportunistic behaviour rise in free markets faster than in cooperations, and much faster than in hierarchies (PIES 2001:9–18; ZENGER et al. 2002; WILLIAMSON 1998:38). In hierarchies, disputes can be settled without the time and costs of using the legal system. There are other transaction costs, however, namely problems of information flow, incentives, bureaucracy, monitoring and performance evaluation (KLEIN 2005:436; MENARD 2005:291).

It is important to understand the economic organisation of interactions between asset specificity and uncertainty. Generally, it can be assumed that if assets are not specific, it is less costly for firms to rely on markets, even if they work in an uncertain environment. If assets are specific and uncertainty is low due to market changes and behaviour of firms being predictable, firms are encouraged to shift from hierarchical to equity cooperation (SHELANSKI and KLEIN 1995:339–340; MENARD 2005:285; WILLIAMSON 1987:60).

The third attribute, **frequency**, proved to be more difficult to enter into operation. The frequency of a transaction is important, but depends on specificity and uncertainty of transactions. For a one-off or occasional transaction, markets seem to be the best solution, but they are also associated with the highest risk, because the trading partners have no economic interest in future relations and may behave opportunistically. Only in the case of less investment in specific assets and low market and behavioural uncertainty do markets become the appropriate option. Transactions involving specific investments are rarely organised via markets alone.

Frequent transactions are organised better in hierarchies or cooperative relations, as these are the best ways to learn from each other and ensure satisfaction (MENARD 2005:285).

2.4.4. Institutional Environment and Governance Modes

Besides the transactional specificity, the institutional setting also determines the transaction costs of different governance modes. Institutional frameworks can provide well developed market institutions supporting interorganisational trade, but they can also be poorly developed, which facilitates control and protection of business in hierarchies. Markets require a dense web of institutions in order to support their existence and development. If a region does not provide appropriate formal institutions to protect transaction on markets, firms tend to organise their relations in hybrid forms, which gives them more certainty for transactions. In hybrid forms, market failures can be compensated for by informal institutions. They take over the functionality of formal rules (CAI and YANG 2008). If informal institutions do not provide a framework for safe transactions, firms organise production in hierarchies, because hierarchies have their own security mechanisms to guard against market and behavioural uncertainty (ZENGER et al. 2002). Moreover, the dysfunctionality of the legal system does not affect hierarchical relations as much, because contracts are not necessary and conflicts are usually settled in-house. But as exchanges with the outside world are limited, hierarchies run the risk of miscalculation or slower adaptation.

When regions undergo times of institutional change, they have to deal with a steadily changing but improving market environment. Against the background of the development in China, institutional change should be contrasted with the choice of governance mode, while fixing specification of the attributes in transactions which determine governance modes. While assuming that the market and behavioural uncertainty remains unchanged, the influence of the institutional setting on governance modes is illustrated here. If the institutional environment improves and provides more safeguards, governance modes start to shift (PIES 2001:17). Whereas transactions with the market and behavioural uncertainty $k > k_1$ are governed by non-equity relationships in an environment with poorly developed safeguards and enforcement mechanisms, in regions providing well-developed safeguards, the same transactions would be organised via markets (see Figure 2.8). The new trade-off point between markets and non-equity relations k'_1 is associated with a higher uncertainty in transactions. All trade-off points behave similarly and shift right ($k_2 \rightarrow k'_2, k_3 \rightarrow k'_3$).

As argued in Section 2.3.1, it takes some time before firms adapt to the new formal institutions. Firms with their own production plants will not divest their plant immediately, but will stay with their former business solution for a certain time. In the case of expansion or completely changing production lines, firms tend to adapt to the new institutional environment. They do not set up new production plants as they would have done before, but look instead for a JV partner or a dense cooperation. They begin to shift from hierarchies to hybrids. Firms safeguard business by relying on newly introduced formal institutions, but they can also revert to informal institutions. Furthermore, hybrid modes give firms the choice between more market-like and more hierarchical-like methods. This results in a greater firm flexibility. In contrast, firms engaged in non-equity cooperation do not switch to markets immediately (shift: $k_1 \rightarrow k_1'$). Trust and safety in their functional business networks keep them away from an immediate shift from cooperation to markets. Therefore, it can be expected that more hybrid governance modes occur during the transformation process (WILLIAMSON 1998; PENG and ZHOU 2005), as the shift from hierarchies to hybrid starts early in the process, whereas the shift from hybrids to markets starts later. The graph connecting all trade-off points indicates the minimal transaction costs. If the trade-off points shift right, the graph of minimal transaction costs moves closer to the x-axis. This shift results in lower transaction costs overall in a region, because they can benefit from the lower costs of markets for more transactions. To sum up, it can be expected that a more developed institutional framework leads to an increased likelihood that transactions with a certain specificity can be observed in a less hierarchical mode of governance, in turn leading to lower transaction costs. It can also be expected that more relationships are organised in hybrid modes of governance during times of institutional change.

2.4.5. Interim Conclusion and Hypotheses derived from Section 2.4

C *How does the institutional environment and the transactional specificity affect the choice of governance modes? What governance modes can be expected in HK and the PRD?*

To answer this research question, transaction cost theory was introduced. It is based on the assumption that markets are not, as neoclassical theory predicts, necessarily the most efficient governance mode to organise customer producer relations (COASE 1937). The use of different governance modes—market, hybrids and hierarchies—is associated with transaction costs for information about potential customers and producers, costs for negotiation, draft and conclusion of contracts and costs to enforce contracts. These costs differ for the three governance modes depending on the institutional environment and the transactional specificity (WILLIAMSON 1987). In a well developed institutional setting, minimal transaction costs are expected on markets, whereas hierarchies are associated with high costs. If the institutional environment is poorly developed, hierarchies provide the lowest costs and the best safeguards for firms against uncertainties. Furthermore, firms tend to look for better safeguards in hierarchies when more transaction-specific investments are involved. When institutional change in transitional economies takes

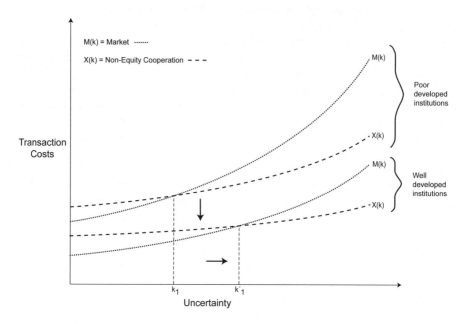

Figure 2.8.: Shift of governance modes according to changes in the institutional environment

Source: Own figure

place, firms act in a setting of permanently improving formal institutions. Firms tend to detach from hierarchical modes and prefer hybrid modes in the first stage. In a second stage, when formal institutions are completely developed, firms should move to market relationships. However, it is also argued that the flexibility offered by hybrids remains attractive for firms. Consequently, they do not switch to markets. In order to differentiate between types of hybrids, they will be divided into categories of equity and non-equity cooperation.

The main aim of this paper is to explain the governance of customer and producer relations of electronics firms in the GPRD by way of institutional factors. With the beginning of the opening policy in China in 1979, HK firms started to move their production plants to the Chinese mainland. At the same time, the demand for consumer electronics grew and Western LFs decided to outsource production activities to lower-cost economies. Electronics firms in HK benefited by providing a good deal: legal security for contract conclusion and enforcement in HK, and cost-saving production in China. HK firms act as intermediaries between global customers and local Chinese producers in the electronics industry (MEYER et al. 2009). HK firms are expected to adapt to the distinct institutional settings which depend on the location of partners. It is assumed that relationships with global partners are organised by markets or non-equity cooperation, as safeguards provided by formal rules guard against opportunistic behaviour. In contrast, it is assumed that HK firms' relationships to China are focussed on strong hierarchies or equity cooperation due to the incomplete institutional setting as source of uncertainty. Moreover, HK firms rely on supply assurance, which can best be realised in hierarchies and relationships of control (BOLTON and WHINSTON 1993). As China is on its way to improving formal rules and stabilising the business environment, relationships set up recently are expected to be less hierarchically organised than those set up around the time that China started its economic opening. Although formal institutions are on their way to adapting to international standards, informal institutions retained their relevance despite being tailored to new contexts. Therefore, relationships to Chinese-based firms are expected to be more frequently organised in hybrid forms.

To prove propositions derived from the conceptual discussion, the following empirical research question and hypotheses were formulated:

I *Do HK firms organise their business relations to customers and producers in the PRD differently from their relations to customers and producers abroad? Is this a consequence of the different development stages of the institutional environment? Is there a trend toward hybrid forms of governance in China resulting from the transitional change?*

Hypothesis C1 HK firms are more likely to govern relationships to global customers and producers via the markets, whereas they are likely to govern relationships to local Chinese customers and producers in a hierarchical way.

Hypothesis C2 Differences in the applied governance modes of HK firms can be explained by the distinct institutional settings in the PRD and HK.

Hypothesis C3 With the progress of institutional change in China, HK firms' relationships to PRD-based firms have shifted from strongly hierarchical organisations to hybrid modes, and lately to more market-like relationships.

2.5. FORMAL AND INFORMAL INTERACTIONS WITHIN CUSTOMER PRODUCER RELATIONS: A FIRM'S PERSPECTIVE

The previous section focussed on the institutional environments and the different governance modes they foster. This section goes beyond the selection of governance modes. It will outline the interplay of formal and informal interactions within customer producer relations, while looking at *recruitment procedures, contractual arrangements, enforcement mechanisms*. According to the model of LEWIN et al. (1999), the interaction processes will be analysed against the background of institutional and industrial dynamics. Answers will be provided for the following (research question **D**):

- under what circumstances do firms prefer informal to formal practices for organising their customer producer relations?

- how does the use of informal interactions in recruitment procedures, contractual arrangements and enforcement mechanisms affect the flexibility of firms?

- what differences are expected between the behaviour of firms in HK and the PRD?

To pursue their business interests, firms need transactional safeguards. Section 2.5.1 focuses on the variety of recruitment procedures for new customers and producers, including different ex ante safeguards for transactional success. Those safeguards are designed to increase the likelihood of conflict-free trade while also choosing the right partner. Procedures will be divided into formal and informal practices. The way contractual arrangements are concluded provides additional safeguards (see Section 2.5.2). Section 2.5.3 concentrates on the ex post enforcement mechanisms which come into force when ex ante safeguards are not sufficient to guarantee smooth trade and when disputes need to be settled.

This is the foundation for a systematic study of the interaction processes of firms in China and HK which is designed to illustrate the differences caused by the institutional environment. It is a response to the request of COASE (1992:719) for empirical data on contracts to support the theoretical discussion.

> "I have come to the conclusion that the main obstacle faced by researchers in industrial organization is the lack of data on contracts and the activities of firms"

2.5.1. Contacting and Selecting Customers and Producers

In this section, the recruitment procedures for customers and producers are examined. Recruitment processes are divided firstly according to the channels used to contact customers and producers, and secondly according to the selection process. In the neoclassical theory, firms are expected to select their business producers and customers by evaluating their rational factors of success (for example firm performance, product quality etc.). This could be interpreted as an impersonal and therefore formal method of recruiting customers and producers. But new approaches add softer evaluation criteria to this collection (for example personal relations, industry experience). These contain more personal and therefore informal elements. GRANOVETTER (1985) is thought to be the first person to have intensively studied the effect of *social networks* on business success. He found that social networks are transformed into business networks. He developed the concept of "embeddedness". This implies that the structure of social networks determines the choice of economic trading partners and how to interact with them. Empirical information on social networks proved positive effects on economic development and democracy (PUTNAM

1993; PUTNAM 2000), economic exchange (COLEMAN 1990; BURT 1992), education (COLEMAN 1988) and career success (BAKER 2000). Social networks are
particularly beneficial when economic transactions take place in an environment of
high risk and uncertainty. Elaborating on social networks provides an insight into
why social networks are beneficial for managers. An examination of the Chinese
business culture and their guanxi networks strengthens the assumption that personal
networks have a positive influence on a firm's flexibility. When managers decide
to use their personal networks to recruit new customers and producers, for example, it usually takes less time to contact them than using an official bidding process.
The informal character of interactions in social networks and its advantages for the
organisation of customer producer relationships will become clear. According to
the definition of informality in Section 2.3.3, the recruitment process via social networks touches on the dimension of personalisation. Recruitment can be based on
personal or impersonal ties, which refers to informal or formal modes respectively.
The way firms choose to recruit customers and producers provides the basis for the
level of informality applied in the two other dimensions—contractual arrangements
and their enforcement.

Guanxi—Chinese Business Networks

In China, business and social networks are not separated, but combined to produce
a mixed form of network, which is called *guanxi*. Guanxi indicates a personalised
network of influence. Basically, guanxi describes a personal connection between
two people or a network of contacts, the idea of which is to draw on established
connections in order to secure favours in personal interests (DUNNING and KIM
2007; LUO and CHEN 1997; PARK and LUO 2001). A manager's guanxi network
is part of the Chinese way of doing business. It can be seen as a special form of
business network which has strong affiliations with social networks. The use of
guanxi networks is deeply rooted in Chinese society, as opposed to an adaptation to
the political structure. But it should be mentioned that guanxi is an instrument with
special traits which are used to fill gaps in the formal institutional setting.

The traits of guanxi are manifold (see Table 2.4). Firstly, guanxi operates on the
level of individuals and is highly *personal*. Psychological research has shown that
in personal contacts, face-to-face visual clues, such as facial expression, gesture and
posture play a significant role in understanding and trusting the other party and give
all participants the feeling that the other can be influenced more, thus making it appear more worthwhile to enter into cooperation (PUTNAM 1993; MISZTAL 2000).
MISZTAL (2000) proved that informal meetings still matter, as face-to-face contacts
have the potential to draw individuals deeper into relationships with one another
and to develop trust and intimacy. MACAULAY (1963) observed that businessmen
still trust in and rely on a man's word in a brief letter or a handshake, even if they
take higher risks concerning the transaction. Moreover, experiments revealed that

Table 2.4.: The traits of guanxi

Traits	Description
Personal	Guanxi is established between individuals
Reciprocal	An individual's reputation is tied up with reciprocal obligations
Intangible	Guanxi maintained by an unspoken commitment
Long-term	Guanxi is reinforced through long-term cultivation
Transferable	Guanxi is transferable through a third party as a referral
Utilitarian	Guanxi is purposefully driven by personal interests

Source: According to DUNNING and KIM (2007:331)

in face-to-face discussion, participants reach agreement sooner than in any other negotiation setting. This leads to greater confidence in the negotiation processes. Additionally, studies discovered that personal contacts make negotiations more of an issue between two people than between two firms, and negotiations are less formalised. Secondly, guanxi is *reciprocal* in the exchange of favours. Reciprocal processes are seen as a supplement to formal procedural rules. They involve mutual expectations on the part of the economic agents that a favour granted now will be repaid later. It is perpetual. Reciprocity can also result in mutual dependency, which might limit the flexibility to choose between different options. This can negatively effect the relationship between informality and flexibility. Thirdly, guanxi is an *intangible* asset. People in the network are committed to each other by an informal and unwritten code of trust. This guards against opportunistic behaviour, as the reputations of individuals are damaged when they disappoint their partners. Fourthly, guanxi is *long-term oriented*. Every guanxi relationship expands the stock of relational capital. It takes time to establish a functional guanxi network. Investments are initially high, but if a guanxi network is established it enables firms to react quickly to changes and provides them with a certain degree of flexibility. Fifthly, guanxi is *transferable*. If *A* has guanxi with *B*, and *B* is a friend of *C*, then B can introduce and recommend *A* to *C*. Finally, guanxi is a *utility* concept. It is based on the exchange of favours, not on sentiments. If guanxi with other parties is no longer achieving objectives, it is easily broken (DUNNING and KIM 2007:330–331; LUO and CHEN 1997; PARK and LUO 2001:457).

In regions such as China, which prosper based on those social networks, the players usually see their mutual confidence as a natural fact rather than an adaptation to the political structure. Therefore, guanxi relationships cannot easily be created— they have their roots in culture and history. FEARON and LAITIN (1996:730) found that social networks are better developed and interactions are more frequent within ethnic groups, because they share certain similarities. Guanxi provides a good example of that. Therefore, social networks are not necessarily related to the spatial proximity of players, although in many cases they are (SABEL 1993:1133; DE MESQUITA and STEPHENSON 2006:43–44).

When contrasting the guanxi network in China with business networks in developed countries, significant differences become apparent. Firstly, the two networks suggest *different steps and methods* of doing business. In China, one builds the relationship first and the transactions will follow. A signed contract marks the end of the first stage in business dealings. But contracts are not necessarily seen as binding - they are subject to change. People abide by the content of contracts simply because of their high personal regard for their business partner. In Western business culture, transactions are put first. Relationships are the second stage and are sometimes seen as consequences of transactions. For Westerners, keeping promises results from their respect for the contract, not for the trading partner. Secondly, the Chinese are more people-oriented, while Westerners are more task-oriented. They focus on *different factors*. The Chinese are more concerned about the harmony of the group, because they believe that respect and caring for others results in future business advantages. This situation makes business difficult between Westerners and the Chinese. As AI (2006:107) put it, "a Westerner would say that the Chinese cannot be trusted because they will always help their friends, and a Chinese would think that he would never trust Westerners since they would not even help a friend." Thirdly, the *scope of perspectives* is different. Westerners tend to focus on the smallest possible unit of economic exchange—a single transaction—in the shortest time. In guanxi networks, however, businessmen focus on long-term alliances, which enables them to react quickly (AI 2006).

Advantages of Contacting and Selecting within Networks

When firms want to **recruit** new customers and producers, they can choose from a variety of methods ranging from very impersonal to very personal. Whereas a bidding competition or an internet search can be deemed highly impersonal and an exhibition moderately personal, recruitment through networks is highly personal. Using networks, firms can either transact with customers and producers they know, or they can use the transferability of networks and interact with a third party (DIMAGGIO and LOUCH 1998). When firms in China use their guanxi networks, they profit from the intangible, transferable and reciprocal character of the network. Other network members are committed either to providing their services to the firm (intangible), or they recommend a third party (transferable). Firms save ex ante transaction costs for screening their environment for new customers and producers, collecting information and establishing the first contact (POWELL 1990; DIMAGGIO and LOUCH 1998; GRANOVETTER 1985). As networks are reciprocal, firms which profit from other members are willing to repay favours. Preliminary empirical evidence has been found in the study of GASTON and BELL (1988). They proved that two-thirds of investors in the informal capital market find investments through their networks. The most reliable sources are friends and business partners. SHILLER and POUND (1989) confirmed those findings. They found that the majority of purchasers of publicly traded common stocks buy based on information from friends and business associates. Personal relations support the process to gain information about the expectation of others (CAI and YANG 2008). Here, the issue of multiple levels in networks must be discussed. Networks are usually developed on an individual level, but are often transferred to a firm level. This does not skew empirical finding as long as firms are small and/or family-led. The owner and manager are the same person and the owner's network is identical to the firm's network. This mainly applies to SMEs in HK and China. In large firms with frequent changes in management positions, a distinction must be made between the networks of individuals and those of the firm. This is also the case for large CMs in HK.

When firms can revert to networks, they enlarge their pool of choices for recruiting new customers and producers. Very impersonal contact channels are at everybody's disposal, but personal channels are not available to all firms. They can be seen as additional opportunities to use. Moreover, under volatile industry and market conditions with changing demand, time is essential for a firm's competitiveness. As a personal network is always ready to use, it seems to be a quicker way of activating new customers and producers at short notice (MILLINGTON et al. 2006). It can be concluded that personal contact channels (informal dimension) provide an additional option for firms to contact potential partners quickly and cost-effectively. As speed is demanded in response to changing markets, a network helps to identify who has surplus orders and whose machines are available. Networks smooth the flow of information and resources between firms and the order-to-delivery cycle time shortens (LEE 1997:194-195). This increases the flexibility of firms (VOLBERDA 1996:361). Empirically, MENARD (2004:361) could prove that the use of informal modes through networking to recruit new customers and producers opens up additional opportunities for firms without weakening any formal modes. Nevertheless, networks can also lead to a lock-in if firms rely on them exclusively.

Once contact channels are established, potential customers and producers have to be **evaluated and selected by firms**. Again, the neoclassical literature assumes a rational selection process (ANGELES and NATH 2000; HITT et al. 2000; SAFFU and MAMMAN 2000). But bounded rationality limits the scope of firms to assess a potential partner's quality and reliability in advance. Firms can decide either to try to assess potential partners quality and reliability within a screening scheme of their own (formal), or they can trust in a network of firms with pooled experiences (informal). Researchers such as DUYSTERS et al. (1999) and recently PIDDUCK (2006:265) argue "that a new perspective on partnership is needed to handle recent rapid economic and technological developments, and propose a more personal approach to partnership". As guanxi networks in China provide a personal base, intangible assets and a long-term perspective, firms can benefit from ex ante safe-guards concerning their choice of new partners. In long-term personal relationships, trust is built on the performance of other parties. Former experiences can be used as an indicator for future satisfaction (MACLEOD 2006). If a firm has no experi-ences of its own to evaluate, the pooled experiences of the guanxi network can be used to investigate reputations. If a producer has acquired a good reputation, it is treated as an asset which loses its value should the producer disappoint customers. In case of a breach, the disappointed party carries out actions that harm the repu-tation of the breaching party. Producers would lose their reputation and customers would refuse to continue buying, believing that further breaches are likely in the future. Firms trade off the benefit of breaching against the cost of damaging their reputation (MACLEOD 2006:6; ARAUJO and ORNELAS 2007:3). Using the char-acteristics of networks, firms are able to safeguard the success of their transactions ex ante. The selected customers and producers are more committed to fulfilling a firm's requirements, because they fear losing their reputation otherwise. This re-duces the risk of opportunistic behaviour. Especially when the formal institutional setting does not provide appropriate mechanisms to safeguard future transactions by way of precisely written contracts (formal safeguards), firms have to ensure their business success via the selection process of customers and producers (informal safeguards). As for the contact channels, networks also provide an additional op-portunity for firms to select new customers and producers. Besides the saving of transaction costs for long screening processes and transactional ex ante safeguards, firms can speed up their selection process. Again, the use of personal networks enhances the flexibility of firms (SEABRIGHT 2004).

Special Setting in Hong Kong and the PRD

As the institutional setting in the PRD is still in transition, Chinese firms are ex-pected not to rely on formal safeguards alone to ensure transactional success. The reasons for this lie in the dysfunctionality of the formal safeguard system. Even though the Chinese government has improved the legal system and encourages firms to make use of Western-style contracts, firms cannot rely on the Chinese court sys-tem to enforce contracts in case of conflicts. Therefore, Chinese firms must put great emphasis on the careful recruitment of producers and customers as a safe-guard for their transactions (MILLINGTON et al. 2006). Bringing the recruitment process to a personal level and therefore using an informal practice, firms hope to achieve better safeguards than in a formal process. Moreover, firms benefit from a quicker recruitment process via networks. However, they also have to deal with the disadvantages of networks, which can also repress and hinder business opera-tions. HUMPHREY and ASHFORTH (2000) showed that interpersonal relationships can also lead customers to award contracts to producers who have higher unit costs, lower quality and slower delivery times. Therefore, they need to be supported by formal structures to guard against the negative effects of informality, such as opportunistic behaviour and its intransparency (MISZTAL 2000:7). In developed institutional environments such as HK, firms are expected to take into account formal and informal recruitment methods and balance them. Although formal safeguards can be ensured by a functional legal system and transactions with strangers can be ensured by contracts, recruitment within networks might still be appealing because of cost and time-reducing aspects. This raises the question of how guanxi networks are affected by institutional change. XIN and PEARCE (1996) and NEE (1992) see guanxi as a substitute for the deficiencies of formal institutional artifacts and failures in the legal system. Proponents of this thesis argue that guanxi will become less important in the future. On the other hand, other experts argue that guanxi is also transforming into a more functional business network. Whereas the traditional guanxi network consists of family (zijiaren) and fellows/helpers (shouren), it has now changed to strangers/business partners (shengren), which makes it more effective for business (MILLINGTON et al. 2006). Studies of business culture in HK or Taiwan are widely used to provide examples of the transition—but nevertheless survival—of guanxi in the modern market (FAN 2002; ZHANG and ZHANG 2006). Therefore, HK firms are expected to recruit customers and producers in China within guanxi networks, whereas they might apply a mixture of varieties to recruit customers and producers in Western countries.

2.5.2. Concluding Contractual Arrangements

When a business partner is selected, contractual arrangements have to be discussed. A variety of contractual agreements ranging from formal to informal provide different safeguards for transactions. Principle agent theory and contract theory discuss the trade-off between formal and informal contracting against the background of the institutional environment.

The contract theory focuses on a group of firms economising on joint surplus. Contractual arrangements are needed to direct the economic exchange. If actions for every possible state of the world could be specified in a legally enforceable contract—called *complete contract*—the organisation of economic exchange would be easy and safe. Parties would design a contract covering all contingencies, future possibilities for deals and the reaction of both parties to it. Behaviour deviating from the agreements in the contract would result in penalties which could be enforced by courts (MASTEN 2000). But in the real world, economic agents firstly behave opportunistically, and secondly act according to bounded rationality as outlined in the assumptions of the NIE (see Section 2.3.1). This leads to (1) information asymmetry between customers (principle) and producers (agents), and (2) to incomplete contracts.

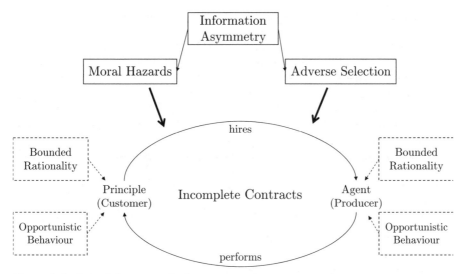

Figure 2.9.: Principle agent relations
Source: Own draft

(1) Principle Agent Theory

The *principle-agent-theory* assumes that the principle (customer) entrusts the agent (producer) with a job. He uses the agent to follow his own goals. The principle expects the agent to put his entire effort into fulfilling the principle's job, rather than having his own goals in mind. But the principle only has limited opportunities to monitor and control the agent's engagement (occurrence of hidden actions), because of his bounded rationality (see Figure 2.9). He only sees the result of his effort (for example the firm's sales). It can be assumed that the principle and agent have individual interests and different aims. Therefore, the agent acts with the advantage of information, because he knows about both his own interests and those of the principle. He can use information asymmetry to the principle's disadvantage (opportunistic behaviour). The fact that customers have to deal with a lack of information results in two issues typically characterising principle-agent-relationships: moral hazards and adverse selection (MCMILLAN 1995:205).

- **Moral hazards:** Customers fear moral hazards when agents (producer) deviate from joint surplus maximising behaviour after an apparent omission of risk. For example, it is possible that a customer urgently searching for a specialised producer may subcontract to that producer before the producer realises the rarity of his production capabilities. The producer's awareness of this rarity reduces the risk of losing the customer. He would then start to strengthen his position against the customer. Customers fear moral hazards stemming from the information available to producers.

- **Adverse Selection:** Adverse selection indicates a situation in which principles (customers) can only choose between agents (producers) with less desirable characteristics. For example, a customer is looking for a producer, but has no information about the producer's quality. Therefore, he is only willing to pay a medium price (middle-of-the-road-choice) for the producer's products. However, excellent producers are less willing to sell their products for a medium price, whereas producers with a poor quality are more willing. This leads to poor quality products on markets, which is less desirable (MASTEN 2000).

A principle can reduce his risks by selecting the agent carefully. Efficient means of recruitment are described in Section 2.5.1.

(2) Incomplete Contracts

There are two reasons for the existence of incomplete contracts. Firstly, it has been shown that bounded rationality and opportunistic behaviour on the part of principles and agents results in information asymmetry. As complete contracts require complete information, it must be concluded that contracts cannot be completed, but rather always remain incomplete (TIROLE 1999:741; SEGAL 1999:57; HART and MOORE 1999:116). The more complex the deal, the more difficult it is for parties to be able to provide a clear and enforceable contract in advance which can be verified by courts. A second, but weaker reason for contractual incompleteness is that transactors leave contingencies unfixed when the costs of anticipating, devising optimal responses to and drafting provisions for future events are higher than the expected gains from doing it. This is a trade-off between the marginal cost of designing contingencies in a contract and the marginal utility of gains and risk coverage. As there is no commonly agreed definition for the incompleteness of contracts in the literature (TIROLE 1999:743), this work refers to contractual incompleteness according to HART and MOORE (1999:134). Firms

> "rather [...] write a contract that is *incomplete*, in the sense that it contains gaps or missing provisions; that is, the contract will specify some actions the parties must take but not others; it will mention what should happen in some states of the world, but not in others. (HART 1988:123)"

MASTEN (2000:29–31) distinguishes between incomplete contract models with and without renegotiation (dynamic and static models). In static models, the contract theory assumes that parties agree not to renegotiate the contract even if a Pareto improvement would be possible for both parties. In dynamic models, "[...] parties agree some things now and agree to agree other things later" (HART and MOORE

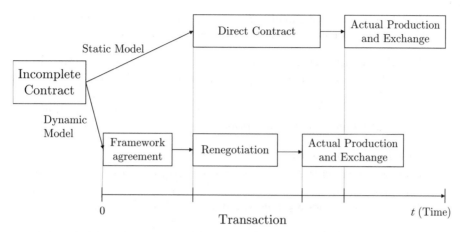

Figure 2.10.: Static and dynamic contracting model

Source: Own draft

2004:2). This work follows the suggestion of MASTEN. It distinguishes the dynamic from the static model. A dynamic model exists when a framework agreement (FA) is concluded in advance with an opportunity for renegotiation (RN). A static model exists when a direct contract (DC) is concluded containing all contractual details without the possibility of renegotiation (see Figure 2.10).

- **Framework agreements (FA) with a renegotiation (RN) process:** For some economic exchanges, the contractual parties decide to conclude a FA in advance and renegotiate or finalise details when an order is placed (MENARD 2004:352). For example, suppose that a customer and a producer sign a contract to trade DVD players a year in advance. They may decide now on the price and on the broad product specification, but may agree that the exact number of items and the time of delivery will be determined later, when the actual order is placed (MASTEN 2000:37). Moreover, FAs often couple a customer's minimum commitment to purchase a certain percentage (e. g. 10%) below the forecast with the supplier's guarantee to deliver a certain percentage (e. g. 5%) above if requested. FAs often include constraints on the frequency of quantity changes by the customer (KRAJEWSKI et al. 2005:454). As this requires a critical amount of commitment and long-term orientation between parties, this process is expected to take place in non-equity cooperation, equity cooperation or hierarchies, but not in markets. If no equity is involved in the relationship between customers and producers, FAs are the only indicator distinguishing non-equity cooperation from markets. If equity is involved (equity cooperation, hierarchies), FAs might also be used to regulate transactions between parent companies and their 100%-owned subsidaries, for example, or parent companies and their associated JVs.

- **Direct contracts (DC) without renegotiation:** Contractual parties decide not to have a supporting FA in advance, but settle all details directly with the placement of orders. DC and the RN process within FAs take place at the same time—when an order is placed. In RNs, only final details need to be agreed on, whereas DCs involve the entire negotiation process being finalised with the placement of orders. Therefore, it is expected that DCs need more time than RN. Markets are defined by the DC model. Firms in equity cooperation and hierarchies which strongly control their affiliates are expected to settle transactional information directly (DC), whereas when they have less control they are expected to use the FA and RN.

According to the definition of informality in Section 2.3.3, contractual arrangements touch on the dimension of precision (precise (formal) to open-ended (informal)) and form of contracts (written (formal) to verbal (informal)). The precision of contracts can be measured by the level of details fixed. The overall recommendation of the minimum of details fixed in contracts includes: quantity, quality, specification and price of products as well as a clause stating what to undertake in case of conflicts (WILLIAMSON 2000:602-603). FAs can be divided into very precise agreements, e. g. when there is little or nothing to negotiate later, and open-ended agreements, e. g. when there is much to settle later. The RN contract can again precisely fix the remaining points or leave gaps in contracts. The same phenomenon is observable with direct contracts. Besides the level of details, the form of agreements—written or verbal—is an important means of informality. The more details are agreed on in written form, the more formalised the contract is.

Institutions, Contracts and Special Situation in HK and the PRD

When the formal institutional environment does not provide safeguards for the enforcement of contracts, firms' incentive to design precisely written contracts is low. When formal safeguards are missing, firms have to protect their transactional interests in other ways. In doing so, it can be important how the relationship to the other parties is characterised. A trust-based relationship, in which parties rely on reputation and reciprocity, such as guanxi networks, provides informal safeguards which lead to self-enforcing contracts (MENARD 2004:362). Even if formal safeguards are available, firms still weigh the cost of drawing up those contracts against the value which would be lost in case of conflicts. Only when transaction-specific investments are high is it worth incurring the considerable costs of administering complex and highly detailed contractual arrangements (see model in HART and MOORE (2004:5–8) and MASTEN (2000)). In both cases, agreements must be concluded, but their degree of precision or their form is not pivotal for the ability to enforce them. Under certain circumstances, firms are expected to try to agree only on the necessary details (SEGAL 1999:74). Choosing the informal way might be associated with the lowest costs, and it is more flexible for firms, because not all details have to be fixed at the same time. This means that they can speed up the contractual process and revise and renegotiate on a steady platform in order to adapt to changing demand.

Despite improvements in China, the institutional framework does not provide firms with the appropriate safeguards for protecting their transactions efficiently. Even if contractual details are formally fixed and verifiable, it is still difficult to enforce them via the courts (MENARD 2004:360). Therefore, it can be expected that firms tend to rely on informal safeguards to protect business while using informal contracting—more verbal and open-ended agreements. Additionally, the impact of contracts is low in Chinese business culture. Contracts are not the results of intensive negotiations, but they are seen rather as the beginning of a relationship over a much longer time period (see guanxi). Therefore, the permanent renegotiation of contracts is a process which accompanies the development of the relationship. HK firms act in a well-developed legal setting which provides a third-party enforcement system for contracts. But even then, it can be expected that HK firms use informal contracting modes for some parts of their contracts, depending on external requirements. CMs are expected to be in a position to formalise contracts with international customers. Contracts with Chinese producers are assumed to be less formalised. Moreover, large firms are assumed to work more on a formal contracting basis than SMEs. This fits with the requirement of flexibility of Asian producers. In industries with rapidly shifting technologies and markets, LFs have no way to predict what product specification, capacity and design features are needed in the next period. The high degree of uncertainty has important implications for the reorganisation of special contractual arrangements. Firms need adjustable contracts embedded in networks to ensure the ability to react to changing demand profiles of end customers (ERNST 2004:106). LEE (1997:189) shows that one-third of the HK managers rely on verbal contracts. According to LEE (1997:194), "informality provides flexibility for individual entrepreneurs to deal with accidental changes in the business environment while the formality provides a last resort to handle breached contracts." Li-Ka Shing, HK's most famous businessman, agreed to deals of hundreds of millions of dollars without a written contract (CHOI 1994:677). Moreover, it can be expected that contracts become more formalised with the continuation of the transformation process in China, but that informal aspects retain their importance, as they provide firms with more flexibility in some cases. MISZTAL (2000:3) summarises as follows:

> "Managers' appreciation of informal encounters and tacit understanding shows that, although the process of formalization is the dominant trend in modern social life, informality is the essential element in constructing trust relationships and, therefore, in any cooperative arrangement

aimed at improving the quality of life. Nonetheless, [...], managers would not like to abandon the formal guarantees provided by the contract for the advantages of more personalized arrangements. [...] it can be said that the positive value of informality is only ensured in the context of the process of formalization of individual rights and public rules."

2.5.3. Enforcing Contracts

Although optimal ex ante safeguards for transactions are set (careful selection of partners, contractual arrangements appropriate to the institutional environment), disputes may still arise. Those disputes must be settled ex post. Firms try to enforce their concluded arrangements, thereby choosing between formal and informal practices (MACLEOD 2006; CHOI 1994). According to the dimensions of informality and formality, a formal enforcement refers to a tight enforcement by legal power. In contrast, informal enforcement mechanisms refer to a loose enforcement (willingness to deviate from original interests) by social power. The selection of enforcement mechanisms of contracts depends on the ex ante safeguards selected and the quality of the institutional environment (ARAUJO and ORNELAS 2007). For instance, if a contract is mainly settled informally, it will be difficult to enforce it via courts. Firms must switch to less formal enforcement mechanisms. Furthermore, if a contract is formally specified, but costs and inefficiency are high, firms avoid formal processes of contract enforcement (GOLDSMITH et al. 2006:85).

Litigation is widely recognised as the most formal of enforcement mechanisms. However, the high expenses involved in litigation open up the awareness for **A**lternative **D**ispute **R**esolution (ADR). ADR involves *arbitration, mediation* and *negotiation*. It is also interpreted as **A**ppropriate **D**ispute **R**esolution. ADR is characterised by its use of more informal elements in dispute resolution to find the appropriate way to solve problems while saving time and money. "The question is how one should design the rules to encourage efficient settlement rather than litigation" (MACLEOD 2006:13). Instead of making a court appeal, firms should carefully choose the most appropriate series of enforcement mechanisms. This does not mean that formal and informal mechanisms to settle disputes are mutually exclusive. On the contrary, they complement each other. For example, firms can begin to negotiate or mediate and if no solution is reached, they can opt for arbitration or litigation (MACLEOD 2007). In the following section, the four main dispute resolution mechanisms will be described and compared (see Table 2.5).

Negotiation is the simplest means of settling disputes. TAYLOR (2000) examines a model of the old-boy network, in which conflicts are settled by negotiation. In the Hanseatic League, businessmen preferred to negotiate their conflicts rather than use litigation processes. The choice of informal enforcement processes is often associated with informal contracting, since those agreements are less verifiable by courts. Negotiations mostly indicate the first step of dispute settlement. They only lead to a loose enforcement, as compromises are usually reached. The power used

Table 2.5.: Safeguard and enforcement mechanisms of contracts with customers and producers

		Formal	⟷		Informal
Ex ante Safeguards	Recruitment procedure	rational selection	⟷		recommen-dation
			reputation		
	Contractual arrange-ments	precise, written	⟷		open-ended, oral
Ex post Enforcement	Enforce-ment mechanisms	legal power	⟷		social power
		Litigation	Arbi-tration	Media-tion	Negotiation

Source: Own draft

to enforce results is based on social rather than on legal mechanisms. Therefore, negotiation can be taken as a very informal mode to enforce contracts ex post.

Mediation is a voluntary, non-binding, private dispute resolution process, in which a neutral party helps the parties to reach a negotiated settlement. It is one of the less formal alternatives to litigation. It is most prevalent in the commonwealth legal system, especially in Australia. A mediator must be selected collectively by all participating parties. Professional mediators are provided by special mediation associations, but parties generally have free choice.

At the beginning of the mediation, the mediator explains his/her role, the confidential nature of the proceedings, any ground rules and the procedural steps that will be followed. During the fact-gathering stage, the mediator will begin to define the issues, helping the parties to focus on the issues rather than their positions. A mediation process includes caucuses or private sessions. The mediator is not allowed to pass on private information to the other party without approval. Once a tentative agreement is reached, the mediator clarifies the terms of the agreement and makes sure all parties understand the agreement.

Mediation can take place quickly and frequently with relatively little expense compared to a court settlement. It focuses on the parties' real commercial, emotional and psychological needs, and not just on their legal rights. The mediator steers the parties away from past events and focuses them on what they want to see happen in the future. He/she does not recommend, decide or advise the parties, but only mediates between the parties. Mediation gives the parties an opportunity to participate directly and informally in resolving their own dispute. It is usually voluntary, although participation is sometimes required by contract or by court order. Settlement, however, can never be mandatory. Mediation relies on the will to reach an agreement confidentially. This increases the probability of staying with the agreement reached in the future, even if it is not legally binding. As mediation is confidential, it is a means of settling commercial disputes. Nobody loses face or reputation (HONG KONG INTERNATIONAL ARBITRATION CENTRE (HKIAC) n. d.).

If parties cannot find any solution by themselves, they have the chance to continue in an arbitration process (INTERNATIONAL TRADE CENTRE 2001; GOLDSMITH et al. 2006).

Arbitration is a private process whereby a neutral third party (arbitrator) hears the dispute and makes a decision (INTERNATIONAL TRADE CENTRE 2001:37). In contrast to mediators, arbitrators are allowed to make decisions. Arbitration awards are final and legally binding for the parties, and can only be challenged in very exceptional circumstances. An arbitration award has a status similar to a judgment. Arbitration awards are enforceable through the courts of all states which have signed the New York Convention (1958) (UNITED NATIONS COMMISSION ON INTERNATIONAL TRADE LAW (UNCITRAL) n. d.). When arbitration awards cannot be enforced, either because a country has not signed the New York Convention or it is time and/or cost sensitive, firms avoid arbitration (INTERNATIONAL TRADE CENTRE 2001:49–51). Especially in international commercial disputes, arbitration is more efficient than litigation, because awards can be enforced in other countries much more easily than court decisions. Moreover, parties can avoid the high costs of legal action. The advantages and disadvantages of arbitration are outlined in Table 2.6. Arbitration is typically conducted under rules established by arbitration associations (for example the American Arbitration Association (AAA), China International Economic and Trade Arbitration Commission (CIETEC), Hong Kong International Arbitration Centre (HKIAC), Singapore International Arbitration centre (SIAC), International Chamber of Commerce (ICC)). These rules govern the arbitration process. Parties can select the arbitration centre and the arbitrators. Moreover, parties have to agree on the legal system (for example HK law, Chinese law, Singapore law) which they wish to apply for the arbitration process. The UNITED NATIONS COMMISSION ON INTERNATIONAL TRADE LAW (UNCITRAL) (n. d.) designed a Model Law on International Commercial Arbitration (1985), which was ratified by the largest trading nations. Firms are advised to choose a legal system which has adopted the UNCITRAL model law.

Arbitration only takes place when all parties agree to it. Lawyers recommend fixing an arbitration clause in the contract in advance to avoid conflicts about the arbitration process later on. Arbitration clauses should contain agreements on the selected arbitration centre, place of arbitration, the applicable law, number of arbi-

Table 2.6.: Advantages and disadvantages of arbitration

Advantages	Disadvantages
• Own appointment of arbitrators with the required expertise (e. g. in highly specialised areas) • Speed of resolution and the low cost relative to litigation (but the arbitrators need to be paid) • Confidential treatment of parties, proceedings, problems and awards • Awards are legally binding and easier to enforce in other countries than court judgements (142 countries signed the New York Convention on the Recognition and Enforcement of Foreign Arbitral Awards (1958))	• If the arbitrator is an expert within a specified field, he may not have the requisite expertise when the dispute hinges on difficult points of law • Multi-party disputes before the same arbitral tribunal are impossible • Arbitral tribunal has no power to order consolidation of actions • Each case is decided on its merits and is therefore no guide to future similar cases • If parties agree to solve disputes by arbitration in advance, they give up their right to access the courts

Source: According to UNITED NATIONS COMMISSION ON INTERNATIONAL TRADE LAW (UNCITRAL) (n. d.) and HONG KONG INTERNATIONAL ARBITRATION CENTRE (HKIAC) (n. d.)

trators, time limits and the language of the arbitration (INTERNATIONAL TRADE CENTRE 2001:125–152). Despite the legally binding decision, arbitration is not as formal as litigation, because arbitrators, arbitration centres etc. can be selected by the participating parties. However, the process is more formal than mediation and negotiation.

Litigation refers to the legal procedure of solving commercial disputes. Litigation is a mechanism for tight contract enforcement based on legal proceedings. It can therefore be seen as a very formal enforcement practice. Litigation follows the applied legal system of a country. The functionality of a legal system depends on the nation's willingness and ability to apply formal institutions and their enforcement. Economies in transition rely on steadily changing laws and regulations. In this situation, it is difficult for firms to trust and rely on the legal system. Judges are often partial and less familiar with the new system. Legal judgments are also more difficult to enforce in other countries. The location of the court decision should be chosen carefully, especially in the case of international disputes, because this will be the easiest place to enforce judgements. Furthermore, litigation is often associated with higher costs than ADR.

In China, the legal system and the arbitration clauses are under revision. There is continuing improvement, but international standards have not yet been reached. Moreover, China lacks qualified judges because of the variety of new laws they have to deal with. Most of them are former army employees. They have difficulty handling the multitude of new laws. Arbitrators are better educated, but the arbitration system in China is still not up to the international standards. JOHNSON et al. (2002) found that when the quality of law is low, there is a greater reliance upon informal contracts, which require informal enforcement mechanisms. Comparing the Western and the Chinese business cultures, informal mechanisms are often more convenient for the notion of "losing face" in Asia (CHOI 1994; GOLDSMITH et al. 2006:274, 282–284). DJANKOV et al. (2003) provide evidence of how the quality of law and the cost of litigation vary between countries and legal systems. In countries with a low quality of law, the litigation process is more expensive. Therefore, it can be expected that Chinese firms prefer negotiation and mediation for resolving their disputes and enforcing their contracts. As the law and arbitration system in HK is already well established, it can be assumed that firms use a range of different dispute resolution practices depending on the issue to be resolved. The choice between a range of formal and informal mechanisms opens more opportunities for them, which reduces the time and decreases the cost of resolving disputes (INTER-NATIONAL TRADE CENTRE 2001:49–51).

2.5.4. Interim Conclusion and Hypotheses derived from Section 2.5

D *Under what circumstances do firms prefer informal practices over formal ones to organise their customer/producer relations? How do informal interactions enhance the flexibility of firms? In answering these questions, the focus will be on:*

- *contact and selection procedures for customer and producers*
- *contractual arrangements*
- *enforcement mechanisms*

Concepts based on the NIE illustrate that firms in an incomplete formal institutional environment lean towards informalisation when organising customer producer relations with respect to recruitment procedures, contractual arrangements and enforcement mechanisms. Relying on informal business practices saves them transaction

costs, provides them with better transactional safeguards and makes the relationships more flexible in adapting and reacting to demand changes at short notice.
Particularly in China, traditional business culture is based on personal relationships
(guanxi), which encourage informal interactions among managers. Therefore, it
can be expected that Chinese managers prefer informal interactions with their customers and producers. Proponents of the NIE argue that as soon as the institutional
environment improves, firms switch to more formal behaviour. Other experts argue that informal interactions are not only a substitute for missing or dysfunctional
formal rules, but provide advantages even in complete institutional environments.
They complement and support formal behaviour (GRANOVETTER 1985). The sym-
biosis of both makes firms highly flexible. The distinct setting in HK and the PRD is
perfect for investigating this issue, as the networks of HK firms contain Chinese as
well as Western customers and producers. It can be investigated whether HK firms
prefer informal interactions with customers and producers in China only, or whether
this also applies to Western firms. Preliminary empirical studies provide evidence
that informal practices still remain important to a certain degree. Firms know about
the advantages of flexibility gained from them. By studying three different aspects
of customer producer relations, it will be analysed whether informal behaviour has
the same importance for each aspect. To prove propositions derived from the conceptual discussion, the following empirical research question and hypotheses were
formulated:

J *How high is the degree of informality in customer producer relations of HK-*
based and Chinese-based firms? Are there any differences in the behaviour of
firms between the two locations? Does informal behaviour result in greater
flexibility? In answering these questions, the focus will be on:

  • *contact and selection procedures for customer and producers*

  • *contractual arrangements*

  • *enforcement mechanisms*

Hypothesis D1 Firms in HK apply informal modes to recruit and select producers
and customers, but to a lesser extent than firms in the PRD.

Hypothesis D2 Firms are more flexible and successful in business when using the
full variety of formal and informal practices to recruit customers.

Hypothesis D3 Firms in the GPRD rely on informal contracting modes, but HK
firms' contracts with PRD firms are likely to be more informal than contracts with
firms in the Western world.

Hypothesis D4 Firms using the renegotiable contract model are more flexible than
firms using direct contracts.

Hypothesis D5 Informal contracts enhance the flexibility of firms.

Hypothesis D6 HK firms enhance their flexibility while exploiting the entire spectrum of formal and informal enforcement mechanisms, whereas in the PRD informal practices are most popular.

Hypothesis D7 Small firms, young firms and firms with poor technological skills are more likely to utilise informal enforcement mechanisms.

2.6. DEVELOPING A FRAMEWORK FOR THE EMPIRICAL ANALYSIS

LEWIN et al. (1999) provide a framework in their paper for conceptually studying firms' adaptation processes and the coevolution of industrial and institutional dynamics. It has been shown that leading customers require a high degree of flexibility from Asian-based producers to integrate themselves into global value chains. Flexibility is defined as the variety of options for reacting (qualitative flexibility) and the speed (quantitative flexibility) with which they can be applied. In order to reach flexibility, firms are expected to benefit from using informal means to coordinate their customer producer relations. The degree of informality applied depends on the completeness of the institutional setting. Figure 2.11 summarises the conceptually discussed aspects of firm adaptation (see Chapter 2) from a global, regional and firm perspective on the left, additionally indicating the implications derived for the empirical analysis on the right.

Beginning on the top left hand side, the *industry and market conditions* (see Section 2.2) are conceptually discussed from a *global perspective*. The discussion shows that fragmentation and the global spread of production networks lead to the modularisation of the value chain. Firms in emerging economies take over parts of the module-based production, but LFs require a high degree of qualitative and quantitative flexibility. The empirical part of this work aims to provide evidence for the development of the electronics industry and the modularisation of the value chain. Moreover, it is investigated as to what extent HK and the PRD have benefited from the global growth of the *electronics industry*. It will be established whether manufacturing firms in HK and the PRD are required to work in an unpredictable market environment (see Section 4.2, 4.3 and 4.4). A low predictability of market changes requires a high degree of flexibility of firms if they are to operate successfully on markets.

Informality can be a means of achieving flexibility. Informality in interactions is sometimes a necessity, as formal institutions do not provide sufficient safeguards for transactions. But informality in interactions can also be seen as an opportunity to deal more efficiently with customers or producers. This mainly depends on the *informal and formal institutional environment*, which has been conceptually discussed in Section 2.3 (bottom left hand side of the figure). Here, it is assumed that institutions take effect on a *regional level*. The empirical analysis will show

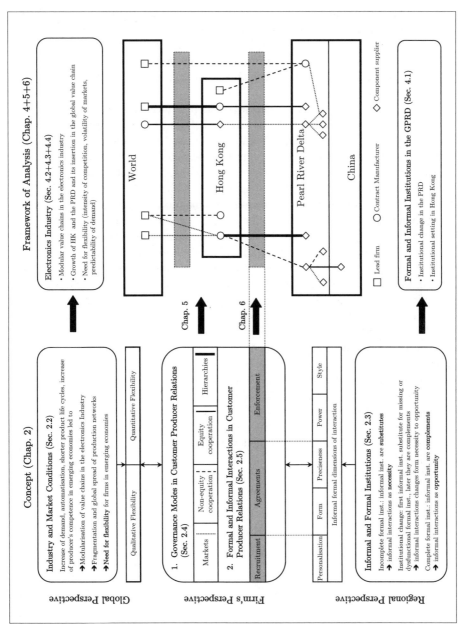

Figure 2.11.: Framework of analysis of customer producer relations

Source: Own draft

whether the institutional setting, which differs in HK and the PRD, is responsible for the informal interactions within customer producer relations, or whether informal interactions are independent of the institutional setting and can be expected to be market-driven. Therefore, the *institutional environment in the GPRD* supporting customer producer relations will be evaluated in Section 4.1 (see bottom right hand side of the figure).

LEWIN et al. argued that the industrial and institutional dynamics influence (1) the *governance modes* (discussed in Section 2.4) and (2) the *formal and informal interactions* in customer producer relations (discussed in Section 2.5) at a *firm level*. The two points are illustrated in the centre of the figure's left hand side.

Firstly, the governance modes represented are markets (dotted line), non-equity cooperation (dashed line), equity cooperation (thin line) and hierarchies (thick line). Initial findings provide a detailed insight into the modes used by firms in HK and the PRD to govern their customer and producer relationships (see Chapter 5). There will be a comparative study which takes into account the governance modes of HK firms to their worldwide customers and their producers in the PRD. Furthermore, the relationships of PRD firms to their customers and producers will be investigated. It will become clear how the industrial and institutional environment influences the selection of governance modes. The corresponding empirical section is depicted in the centre of the figure's right hand side.

Secondly, the formal and informal interactions in the three research areas—recruitment, agreements, enforcement—have been discussed conceptually in Section 2.5 (grey bar in the centre of the figure's left hand side). Here, five dimensions of informality have been identified. The empirical part will focus on the interplay of formal and informal interactions between HK and global firms, and between firms in HK and the PRD. It will be shown under what circumstances firms rely on informal interactions and whether this increases their flexibility. This issue will be addressed in Chapter 6 and is indicated by the grey bars in the centre of the figure's right hand side.

Before empirical results are presented, the next section will discuss how informality and flexibility can be operationalised and how they were measured in HK and the PRD.

3. OPERATIONALISATION, METHODS AND DATA

This work aims to measure how informal interaction of firms with their customers and producers affects the flexibility of firms in terms of variety and speed. Although the importance of informality is recognised in the academic literature, secondary data are rarely available. The combination of informality and flexibility in particular has not been studied systematically until now. Therefore, it was necessary to conduct a study collecting primary data on this relationship. This chapter aims to answer research question **E** and **F** regarding the operationalisation and measurement of informality and flexibility. Section 3.2 concentrates on the operationalisation of informality and flexibility by the identification of indicators. Section 3.3 discusses which qualitative and quantitative methods were applied to collect and analyse data. Emphasis is put on the assessment of the research methods. Firstly though, Section 3.1 briefly describes the selection of the research region and sector. Finally, Section 3.4 analyses the validity of data and the limitations of generalisation.

3.1. SELECTING RESEARCH REGION AND SECTOR

This work is developed within the project "Agile Firm Organisation in the Greater Pearl River Delta", which is in turn one of nine projects in the Priority Programme (SPP) 1233 "Megacities—Megachallenge: Informal Dynamics of Global Change", funded by the German Research Foundation (DFG). The Greater Pearl River Delta (GPRD) was selected as the research region because of the two distinct institutional settings. The GPRD consists of the Hong Kong Special Administrative Region (HK), the Macau Special Administrative Region and the Pearl River Delta Economic Zone (PRD), a part of the Guangdong province (ENRIGHT et al. 2006). In HK, one finds a complete institutional setting, whereas in the PRD, there is ongoing institutional change. The intensive interaction between HK and the PRD has contributed greatly to making the GPRD one of the world's leading manufacturing centres. HK is the centre for management, information, coordination, finance and professional services, while the PRD is the centre for production and operational services. Macau is not considered in this work because its economic power in manufacturing is insignificant compared to HK and the PRD, its economic competitiveness largely resulting from tourism and gambling. The PRD economic zone consists of nine of the 21 cities in the Guangdong province—Guangzhou (GZ), Shenzhen (SZ), Dongguan (DG), Foshan, Zhongshan, Zhuhai, Jiangmen and parts of Huizhou and Zhaoqing. For the purposes of a survey, it is best to remain within particular municiple boundaries (jurisdictions).

Instead of studying the manufacturing industry broadly, the research team decided to concentrate on only one industry—the electronics industry. Reasons for this were:

- the novelty of the research objective (informality achieving flexibility) which requires an explorative approach concentrating on the content

- differences between the structure of industries, because what seems to be very important in one industry may be rated differently in another (the textile industry in China, for example, was long subject to the allocation of quotas, whereas the electronics industry has always been liberalised). Concentrating on one industry also allows for the exclusion of influencing variables other than transaction costs (MENARD 2005:293; SHELANSKI and KLEIN 1995:285)

- the concentration on the largest industry, because it best mirrors the business interactions in which firms are typically involved (see Figure 3.1)

- the electronics industry is subject to the regional division of work (see EN-RIGHT et al. (2005:111))

- the electronics industry is modularised and subject to many interactions along the value chain

- the electronics industry provides researchers with a variety of products ranging from high to low technology, which allows a comparison between single segments of the electronics industry

- the electronics industry is subject to steadily changing demand and pressure to adapt to customers' requirements.

Figure 3.1 illustrates the impact of the electronics industry in different PRD cities. The gross industrial output value (GIO) of the electronics industry in 2005 shows that most cities in the PRD are dominated by the electronics industry. Petroleum and chemistry, textile and garment, food and beverages and building materials only play a minor role. The selection of cities was made on the basis of economic density in terms of the electronics industry. The research team agreed to concentrate on GZ and DG for the PRD survey. GZ was selected as one of the cities with a high density for the electronics industry. It was preferred to Foshan and SZ, which are also hubs of the electronics industry, because of better accessibility. GZ, the capital of GD, is the economic center of the PRD. It is characterised by state-owned and domestic enterprises. In contrast, more foreign firms are located in DG because it was opened

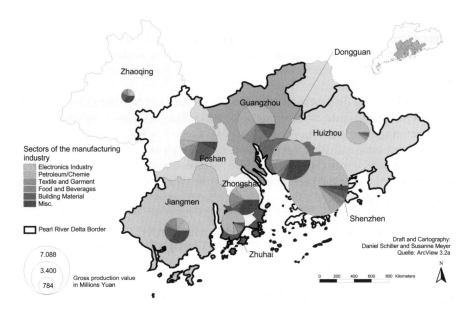

Figure 3.1.: Industrial split of the gross industrial output value in the PRD 2005

Source: Own map according to GUANGDONG PROVINCIAL BUREAU OF STATIS-
TICS (2006)

for foreign investments earlier than GZ. In DG, the economic structure is more
diverse than in GZ. DG was preferred to Jiangmen, which is also characterised
by a mix of industries, because it has a greater absolute power in manufacturing.
Additionally, firms in DG were easier to access for the research team. The research
team thus decided to select two cities: one absolutely dominated by the electronics
industry and one which is characterised by a mix of industries. Moreover, differing
business models were expected in GZ and DG. While DG was one of the first cities
to be industrialised by HK firms in the 1980s (along with SZ), GZ developed its
economic power later involving less investment from HK. The two jurisdictions are
also expected to connect differently to HK.

Because of the dynamic of the electronics industry, which is characterised by
the permanent introduction of new products and changing technological standards,
an objective definition of the industry is difficult to come up with (ERNST 2000:1).
Basically, electronics firms can be divided into firms which produce final products
and those which focus on parts and components. According to HOBDAY (2001:14),
final products in the electronics industry can be divided into the categories of consumer electronics (such as hair dryers, toasters, DVD-players), telecommunications
products (such as mobile phones, landline telephones, routers), computer equipment (such as personal computers, notebooks, hard disks) and electrical machinery

(such as robotics, instruments of measurement and control). The most important components for electronics products are semiconductors. The share of electronics products consumed by other industries, for example the automobile industry, has increased rapidly. The electronics industry is therefore a *key industry*, as other industries profit from it.

A study by the FEDERATION OF HONG KONG INDUSTRIES (FHKI) (2007) has proven the relevance of HK firms in the PRD. They estimated that about half of manufacturing enterprises in the PRD are HK-funded or HK-related firms. These firms employ approximately 9.6 million people. Major investments in the PRD are also made by HK firms. This was the basis for the decision to concentrate on HK-owned and Chinese-owned electronic firms in the GPRD, although multinational or foreign firms were only of minor importance at this explorative stage.

3.2. OPERATIONALISING INFORMALITY AND FLEXIBILITY IN THE ELECTRONICS INDUSTRY

3.2.1. Recording Data on the Need for Flexibility

First of all, the need for flexibility must be measured in order to gauge the pressure exerted on firms to adapt to market changes in the electronics industry. General trends in the electronics industry illustrate its long-term development. As firms in the GPRD depend on the global development of the electronics industry, a *global perspective* has to be applied in order to study the need for flexibility. To gain insights into the special development and growth of electronics firms in the GPRD, however, a *regional perspective* is required to see how firms in the GPRD act under global pressure.

The global pressure to be flexible can be assessed by studying the shorter product life cycles and the high market volatility. This must be contrasted with the proportion of firms in the GPRD which are actually integrated into global value chains. Only these firms have to deal with global pressures. The global impact of electronics firms in the GPRD can be measured by the foreign direct investments (FDI) and the volume of trade and production in the GPRD for the world market. How far-reaching the global pressures are can be seen by the proportion of firms in the GPRD integrated into global value chains. To study the structure of the value chain in the GPRD, the spatial distribution of firms must be analysed. Firms in HK and the PRD were asked to indicate the location of their most important customer and producer to illustrate the network structure (see Chapter 4). The research team decided to focus on the most important customers and producers, as firms are expected to have a large number of both (compare CAI and YANG (2008:61)). The most important producer is defined as the one who produces the largest share of products in terms of value. The most important customer is the one with the highest volume of sales.

From a regional perspective, a growth in the electronics industry and its integration into global networks implies that a larger number of firms are affected by the requirements of flexibility. Besides the global driving forces for flexibility, institutional incompleteness or changes of rules and laws also necessitate corporate flexibility at a regional level.

Secondary data can provide an overview of the global and regional development of the electronics industry. Although electronics firms in the GPRD are affected by the same global and regional forces, the ability of firms to deal with them differs. Therefore, primary data should be collected about a firm's *individual estimation* regarding the intensity of competition and the predictability of industrial changes. If a firm's ability to predict industrial changes is low, that firm will need greater flexibility.

3.2.2. Identifying Indicators for Governance Modes

The governance modes selected to organise customer producer relations can be seen as an adaptation to the need for flexibility. The freedom to select the appropriate governance mode is predetermined by the institutional environment within a region and within a particular industrial dynamic. Firms in HK and the PRD were asked to provide information about what governance modes they have selected to maintain their relationship to their most important customer and producer (see Chapter 5). Governance modes will be divided into market, non-equity cooperation, equity cooperation and hierarchies, as the academic literature suggests. Table 3.1 should give insights into how these terms can be better understood and operationalised in terms of contractual arrangements. Markets are characterised by direct buying and selling contracts without any further agreed cooperation (DC). When firms have agreed to cooperate on a market basis, it is expected that a framework agreement (FA) exists which covers those terms. When an order is placed, final details will be renegotiated (FA+RN). A market relationship changes to cooperation on a non-equity basis. A hierarchical relationship to customers is assumed when a 100%-owned affiliate, subsidiary or parent company places orders with the firm. A hierarchical relationship to producers is assumed when firms have 100%-owned production sites with which they place orders. It is expected that HK firms mainly have their own production sites in the PRD. With respect to contractual relationships, hierarchies can be organised in two ways. Firstly, a hierarchical relationship can be organised via direct contracting (DC), meaning that firms only pass on their orders to their own affiliates within their own company structure via fax, phone call, email or computer-based supply systems. Firms govern, control and monitor the production site completely—this can be referred to as a tight hierarchy. Secondly, when firms consider their (possibly multiple) 100%-owned production sites as independently led firms, they might use framework agreements (FA) outlining what each production site has to achieve. Production sites might compete with each other. This is

Table 3.1.: Relation of governance modes and contracts

Simple classification of governance modes	Contractual Characteristics	Extended classification of governance modes
Hierarchy	Tight Hierarchy	DC
	Loose Hierarchy	FA+RN
Hybrid	Tight Equity Cooperation	DC
	Loose Equity Cooperation	FA+RN
	Non-Equity Cooperation	FA+RN
Market	Pure Market	DC

FA = Framework agreement, RN = Renegotiation, DC = Direct contracting

Source: Own composition

often observed when leading global firms have their own production sites in different countries which compete for production. Hierarchies which use the second contracting model are called loose hierarchies. When firms opt for an equity cooperation with customers and producers, they are assumed to have an equity exchange which is not a 100% equity connection, for example JV or exchange of shares. Again, equity cooperation can be either tightly governed while only using DC or loosely governed while using FA + RN.

In order to establish the determinants of selected governance modes, the location of customers and producers must be taken into account. As a location is always related to a certain institutional environment, it can be investigated whether the institutional environment affects the choice of governance modes, as theoretically expected. The relevance of the institutional environment must be established in comparison to other variables potentially influencing the choice. In order to collect information about the dependency of the customers and producers, which could affect the governance mode, firms were asked to indicate the share of sales or procurement covered by those customers and producers. Additionally, firms were asked to indicate their amount of industry experience (time), allowing an analysis of governance modes in correlation with this factor. Information about other determinants, such as firm size, age, innovativeness, was also collected to measure their influence.

3.2.3. Identifying Indicators for Informality and Flexibility

The research team decided to operationalise and measure informality in contrast to formality. In doing so, the focus was placed on interactions and processes designed by firms or managers. Viewing interactions as a continuum from very informal to very formal gives a better insight into (1) how important informal interactions are in comparison to formal interactions and (2) whether they are seen as the only possible way of organising customer producer relations, or rather as one of several options. The research team decided to ask indirectly, rather than directly, about informal modes applied in customer producer relations. Reasons for this were:

1. the term *informality* is not clear and uniformly defined for everybody

2. *informality* can be sensitive, as it touches on personal impressions and relationships

3. *informality* is associated with "illegality" in HK and China when translated

4. an indirect answer provides additional information about the actual actions of firms in their customer producer relations.

In the following paragraphs, the operationalisation of the three research areas—contact and selection processes for customers and producers, contractual arrangements and enforcement mechanisms—is described with respect to the dimensions of informality and the flexibility implied by different interactions.

Studying informality in *contact and selection processes* for customers and producers, touches on the dimension of *personalisation* (see Table 2.1). Bidding competition is a very formal process. Firms can announce a bidding competition or file a tender to gain new customers or producers. This usually happens over a certain time period and cannot be applied for at short notice. This process is highly impersonal and therefore very formal (see Table 3.2). Contacting new partners on the internet is still impersonal, but a certain preselection can at least be made. This process is not time-dependent and is therefore quicker and more flexible. Exhibitions and fairs as well as sales or sourcing agents work on a formal basis, but they must still win over new customers and producers personally. These procedures are therefore related to some degree of informality. Exhibitions and fairs only take place on certain days each year, which makes this procedure inflexible, whereas sales or sourcing agents can start looking for new customers and producers at short notice, but still need some time until they have successfully contacted them. Former employer-employee relationships are based on personal contacts (informal). They are flexible and can be used at short notice to contact potential customers and producers. Using a network of business or private contacts to recruit customers and producers is highly personal, and therefore informal. These contact channels also provide very quick access to new customers and producers (very flexible).

Table 3.2.: Formal to informal contact channels

Questionnaire	Association with Informality
Bidding competition	very formal
Internet (e. g. e-commerce, direct contact)	↑
Exhibitions and fairs	|
Sales/Sourcing agents	|
Using former employer-employee relationships	|
Business contacts	↓
Private contacts	very informal

Source: Own composition

Moreover, these channels are intrinsically associated with a certain speed which can reflect their quantitative flexibility. Firms also have the option of relying on only one or two contact channels, or of using the entire spectrum, which increases their opportunity to select the appropriate channel to contact customers and producers. A similar approach was chosen for the selection criteria for customers and producers.

Informality in *contractual arrangements* can be measured on two scales: the *degree of precision* and the *form of agreements*. A contract is assumed to be precise when all necessary details recommended by lawyers are covered in the contract. The necessary details involve:

- product specifications

- quantity of products

- price of products

- delivery time

- conflict resolution

The specification of products, quantity, price and delivery time are essential for the production process and supply management. Lawyers strongly recommended including details about conflict resolution, especially when trade becomes inter-national. As two different contract models are considered in this work, it should be established *what* (precision) and *how* (form) details are fixed in those models. Firstly, if firms use FAs, they might fix some points in advance in the FA, while renegotiating or adding details when the actual order is placed. Which details they fix in FAs and which ones they renegotiate or add in the final agreement will be investigated. It is expected that all essential details are covered either in the FA or the RN process. This allows the precision of the FA and the RN to be studied. The fewer details fixed in the FA and the RN respectively, the more informal it is. Secondly, if firms opt for DC, it is assumed that they fix all details when an order is placed. *What* is fixed in agreements entails the question as to *how* it is fixed. The details of contracts can be fixed either in writing (formal), verbally (informal) or both. As orders are often placed at short notice and production is delayed when contract negotiations take a long time, firms must be interested in a quick placement of orders. This means that the time for RN and DC is limited. It should be established how much time firms need to agree on final details when an order is placed. Quick contracting is a response to the required flexibility. It can thus be shown whether informal contracts in terms of precision and form lead to a quicker placement of orders. This would confirm the hypothesis that informality achieves flexibility. Furthermore, the two contracting schemes will be compared taking into account the different governance modes.

The study of *enforcement mechanisms* touches on the two dimensions of informality — *enforcement* and *power*. Enforcement mechanisms range from simple negotiations (very informal), through mediation processes to arbitration and litigation (very formal). To find out whether firms favour informal or formal mechanisms for settling disputes, they were asked to indicate which mechanisms they are willing to use. In terms of the applied flexibility, it must firstly be established how much time certain mechanisms require to settle disputes (quantitative flexibility), and secondly the variety of options firms consider to settle disputes (qualitative flexibility).

3.3. MEASURING INFORMALITY AND FLEXIBILITY IN THE ELECTRONICS INDUSTRY

In line with the project proposal, the research methods had already been specified and could largely be implemented successfully. The research topic required an analysis of primary and secondary data which complement each other. The illustration of the development of the institutional system in HK and the PRD is mainly based on the analysis of secondary literature (see Section 4.1). Additionally, interviews with regional stakeholders were deemed important in understanding the economic environment in HK and the PRD. The global and regional development of the electronics industry is analysed using secondary data (see Section 4.2 and 4.3). Assessing a firm's individual ability to deal with the surrounding conditions required the collection of primary data, as secondary data were not available. As previously mentioned, the idea was to concentrate on the HK and PRD-owned firms, because they have the greatest impact for the economic development of the GPRD. The research team decided to follow the trail of the electronics value chain from the Western world's customers via HK-based CMs to HK-based SMEs, and finally to plants operated in the PRD. This research model enables an analysis of:

- the spatial construction of the value chain in the GPRD

- a comparison of governance modes applied in customer producer relations differentiated according to locations of partners (HK firms' relationships to global and PRD firms)

- the need for flexibility and the informalisation of interactions at different levels of the value chain

In order to study the performance of CMs and SMEs in HK as well as the organisation of electronics manufacturing plants in the PRD, clear research methods were needed. Data about CMs in HK were collected in qualitative interviews, as strategic orientation and firm operations differ, making a standardised survey seem inappropriate. In contrast, SMEs in HK and production-oriented firms in the PRD were surveyed using quantitative methods. They seemed to be more suitable to avoid individual cases being generalised. Although the nature of the study is explorative, it is a basis for further research. The research team agreed that a standardised approach seems to provide a better picture of how firms behave in general. The following paragraphs describe—and more importantly assess—the research methods applied.

1. Analysis of Secondary Data about the Electronics Industry in Hong Kong and the PRD

Data about the electronics industry are available on an international (for example WTO, UNCTAD), national (for example the China Statistical Office) and regional level (for example the Guangdong Statistical Bureau). Secondary data on the activities of firms in the electronics industry must, however, be distinguished from data on product categories. The UNITED NATIONS STATISTICS DEVISION provides an International Standard Industrial Classification (ISIC) for firms' activities and a Standard International Trade Classification (SITC) for product categories. HK and China have only adjusted part of their national classification to the international standards. This implies difficulties in cross-nation analyses or longitudinal analyses (when nations have changed their classification system). However, the research team carefully compared the figures of the different statistical standard classifications and selected those which represent the electronics industry. This allows a comparison of data from different statistical sources (see Appendix A).

Additionally, there is a large and growing difference between the official trade statistics released by China and those released by HK and other countries (MARTIN 2007). SCHINDLER and BECKETT (2005:1) reported that "Hong Kong plays a prominent role as a re-exporter of a large percentage of trade bound for or coming from China. Current reporting practices in China and its trading partners do not fully reflect this role and therefore provide a misleading picture of the origin or ultimate destination of Chinese exports and imports." In this work, an attempt has been made to rely on international sources to report trade data.

When using secondary data from the PRD, one must be aware that it is not fully correct. Population figures reveal the largest discrepancies between the various statistical sources in the PRD. China's statistical approach to population has not kept pace with the actual demographic change. The "registered population" in 2005 was only half of the "interim census" population, as the latter includes migrants. Experts claim that even the "interim census" underestimates the population. Statistically, the recommendation is to work with the census data. The gross domestic product (GDP) in the PRD is measured in accordance with internationally accepted standards and methods, but one has to be aware that firms tend to underreport their sales and earnings in order to avoid taxation. When GDP per capita is calculated, using the registered population should be avoided. Statistics produced in the PRD tend to use these data because an increased GDP per capita suggests economic power. The gross industrial output (GIO) is the volume of industrial products sold in value terms by companies. As the value chain is fragmented, all firms report the value of their products sold, thereby leading to double counting among many enterprises. In contrast, small firms are excluded from the calculation. This leads to an underestimation of the GIO, as the PRD is characterised by SMEs. Moreover, some cities include or exclude certain types of firms, while export processing firms have a special status. There is some evidence to suggest that capital goods associated with export processing do not appear in the statistics. Furthermore, although the system of measuring FDI changed in 2002, many cities in China have not yet adjusted to the new measurement system and many give little indication as to which system they employ. This means that FDI statistics after 2002 differ from earlier ones (ENRIGHT et al. 2006:27–58).

Another difficulty is the definition of the PRD economic zone. Two cities—Huizhou and Zhaoqing—are only partly included. Although secondary data in the official statistics are mainly limited to the administrative border of cities, secondary data in this work include Huizhou and Zhaoqing completely. In case no data on the level of cities are available, this work refers to Guangdong with the knowledge that 80% of the gross value added is concentrated in the PRD (GUANGDONG PROVINCIAL BUREAU OF STATISTICS 2006). To conclude, available secondary data from Chinese statistics will be used in this work due to the lack of other more convenient and trustworthy data, but one has to be aware of their weaknesses.

2. Interviews with Large Electronics Firms in Hong Kong

Method: As the operations of large firms are much more diverse than those of small firms, it was decided that a case study approach is most appropriate to gain information about large players' interaction with their customers and producers.

Selection of firms: For the identification of large electronics firms, different sources were used. Firstly, the HK stock exchange (HONG KONG EXCHANGES AND CLEARING LTD (HKEX) n. d.) was helpful in identifying large listed HK firms in the electronics industry. Secondly, a screening of members of business associations helped to add large non-listed firms to the collection. Thirdly, managers of business associations and organisations rechecked the list of large firms. 25 firms were identified in total.

Implementation: All 25 firms were contacted and asked for an interview with the CEO or one of the senior managers. Some firms could be won through direct contact, but the network approach was generally more successful. Interested and powerful managers of business associations and organisations established the contact to CEOs. Ten firms agreed to a personal interview. Each interview was conducted by between one and three researchers. The rate of return was 40%. The interview followed a guideline which was discussed and agreed on by the entire research team. The interviews were kept to one hour if possible, as HK managers tend to have limited time. All interviews were conducted in English and recorded for a later in-depth analysis. The interviews were prepared in summer 2007 and were conducted between September and November 2007.

Content: The guided interviews covered (i) organisation of production activities in the PRD, (ii) informal aspects in customer producer organisation, and (iii) informal knowledge transfer within innovation activities.

Evaluation: The CEOs of large electronics firms in HK were surprisingly open. Detailed descriptions were given of the organisation of activities and the importance of informal elements to enhance flexibility. The interviews gave a great insight into the strategies of HK firms in coping with uncertainties.

3. Standardised Interviews with SMEs in Hong Kong

Method: The project proposal required 100 firms to be surveyed. The research team decided on standardised interviews, because (1) the standardisation ensures valid and reliable information about the relations between informal interactions and firm flexibility and (2) personal interviews provide the opportunity for interviewees to understand all questions and increase the rate of return.

Selection of firms: For the selection of firms, the research team cooperated with the Research Division of the Trade Development Council (TDC). They provided a database of firms engaged in the electronics sector. According to the HK statistical standard, firms there with more than 100 employees are defined as large, whereas firms with 100 employees or fewer are deemed small and medium sized. The data bank of the TDC contained 4,903 HK SMEs. A random sample of 2,000 firms was the basis for the Social Science Research Centre (SSRC) to conduct the survey.

Implementation: During preparation of the questionnaires in spring 2007, development topics and methods were intensively discussed with the Department of Geography and the Faculty of Business and Economics at Hong Kong University, the Hong Kong Trade Development Council (TDC), the Research Division of the Federation of Hong Kong Industries (FHKI) and the Civic Exchange Research Centre. As native speakers were required for the interviews, the research team decided to authorise the SSRC of Hong Kong University to conduct the interviews. In summer 2007, the questionnaire was completed and translated by a bilingual member of the research team. A letter of recommendation was provided by the TDC. Five interviewers from the SSRC were trained by the bilingual researcher and three pilot interviews were conducted by the trained interviewers accompanied by the bilingual researcher. After contacting the firms to ensure the CEOs' contact details, the SSRC posted the questionnaire to the firms. Afterwards they agreed to an appointment with the CEO for a one-hour interview. The survey ran from August to December 2007. The first 15 interviews were accompanied by the bilingual researcher to ensure data quality. The SSRC completed 104 questionnaires. 90% of the interviews were face-to-face, while 10% of the firms asked for a self-administered version of the questionnaire accompanied by a follow-up telephone interview because the CEOs were spending most of the time in the PRD.

Content: The SME survey in HK addressed the following topics: (i) industry and market conditions, (ii) firm strategy, (iii) organisation of relationships to customers and producers and (iv) organisation of innovation activities.

Evaluation: The SSRC used highly skilled and experienced interviewers for the face-to-face interviews, so the quality of the answers is extraordinarily high. Unfortunately, the original idea of following the links of HK firms to the PRD and identifying their producers for the PRD survey did not work. Firms were reluctant to personally identify their producers in the PRD. The research team had to identify firms in the PRD itself. Industry experience in the GPRD indicates that the timing of the survey should take summer holidays and fair periods in September/October and March/April in HK into account. Additionally, it should be kept in mind that HK CEOs spend much of their time in their PRD factories, which makes it difficult to arrange interviews. For this reason, the research team agreed on self-administered interviews in exceptional cases. To conclude, the face-to-face interviews seem to be a highly appropriate method of collecting data on firms in HK while simultaneously ensuring data quality.

4. Posting to Electronics Firms in the PRD

Method: From an exchange of experiences with partners from the Center for Urban and Regional Studies at the Sun-Yat-Sen University in GZ, it became evident that Chinese managers are reluctant to provide information to researchers. They fear losing their anonymity in personal interviews and tend to give biased information. It was strongly recommended to work with self-administered questionnaires combined with a follow-up telephone interview. As described in Section 3.1, the survey concentrated on the cities of DG and GZ. 53% of the firms in the PRD survey are located in GZ and another 41% are located in DG. In GZ, the Tianhe, Panyu and Huadu districts were most relevant, while in DG the Changan, Dongcheng and Houjie districts were selected.

Selection of firms: In DG, firm data were taken from the Guangdong Electronics Company Catalogue 2007. For the survey in GZ, a list of firms was provided by the Statistical Bureau of GZ. Districts with a high density of electronics companies were identified and firms were randomly selected. Additionally, firms in Zhongshan, SZ, Zhuhai and Foshan were surveyed through acquaintances with the CEO or manager.

Implementation: In a joint workshop with the Center for Urban and Regional Studies at the Sun-Yat-Sen University in Guangzhou, the questionnaire was discussed and students were trained in a two-day session in order to be able to understand the questionnaire and conduct a professional telephone follow-up. Students first rechecked the contact details of CEOs and their willingness to complete the questionnaire by telephone. 300 questionnaires were equally distributed to randomly selected firms in DG and GZ according to the preselection of districts. After their return, students followed up unanswered questions by telephone. In DG, 89 questionnaires were returned (59%). In GZ, 116 of the 150 firms called returned the questionnaire (77%). In the remaining cities, all firms returned the questionnaire. In total, 222 firms participated in the survey. The survey ran from October 2007 to February 2008.

Content: The survey in the PRD addressed the following topics: (i) industry and market conditions, (ii) production system, (iii) employment and (iv) innovation activities.

Evaluation: Difficulties encountered included the high fluctuation of firms, which made it complicated to identify operating from non-operating firms. Some firms also have double registrations. A reliable database of firms is not available and information about the population is virtually nonexistant. Data quality is assumed to be lower than in the HK survey due to less control over the filling in of questionnaires.

5. Interviews with Regional Stakeholders in Hong Kong and the PRD

Method: The position and expertise of regional stakeholders varies greatly, which means that dealing with them requires an individual approach to each case. The research team therefore decided on personal interviews with guided but individually adapted questions.

Selection of stakeholders: The selection of stakeholders was based on content-related specifications. Attention was firstly directed towards firms providing services which contribute to the flexibility of other firms, such as financial or legal service providers. When large orders by international customers require a pre-financing of components, financial service providers become important. Legal advisers, in turn, provide support for the conclusion and enforcement of contracts. The research team then interviewed organisations which support manufacturing firms—business/industry associations and organisations supporting innovation . This provided insights into the general organisation and development of HK and its electronics industry. Finally, academics with experience in researching the HK and PRD regions and persons of authority were selected for an interview.

Implementation: A pool of questions was developed by the research team. The guided interviews contained a selection of questions according to the expertise of the interviewees. The selected stakeholders were contacted directly. The language of the interview was English. 34 interviews were conducted by the research team between September and November 2007, of which 82% were in Hong Kong, with:

- financial service providers (6)
- legal advisers (6)
- business/industry associations (8)
- organisations supporting innovation (6)
- persons of authority (2)
- academics (6)

Content: The interviews provided a comprehensive overview of the historical development of the electronics industry in the GPRD, including the political changes as well as the financial and legal background of the firms.

Evaluation: Regional stakeholders agreed relatively easily to an interview. The overview of the GPRD development provided was detailed and varied, and gives a comprehensive basis for understanding the behaviour of firms. This part of the fieldwork was also designed to arouse the interest of regional stakeholders in the research objectives.

In general, all the surveys conducted in this project are of an explorative nature and an attempt has not been made to produce representative information at this stage. Since the topic of informality defined in an institutional way is rather new, one aim has also been to develop an analytical framework and to test qualitative and quantitative methods of studying informality and flexibility in interactions.

3.4. ASSESSING DATA VALIDITY: COMPARISON BETWEEN SAMPLE AND POPULATION

The sample data of the HK and PRD survey will be used for further quantitative analysis. This section, therefore, compares the sample data with the population data according to the selection criteria to show whether a generalisation is principally possible.

The SMEs in HK were selected according to their affiliation with the electronics industry (see Appendix A) and because of their size of 100 employees or fewer. The TDC in HK provided a databank of 4,903 enterprises fulfilling the selection criteria (HONG KONG TRADE DEVELOPMENT COUNCIL RESEARCH DEPARTMENT (HKTDC) n. d.). Two samples of 1,000 enterprises each were randomly preselected. Of the 2,000 enterprises, 104 were interviewed in HK. The TDC data bank provided a classified number of employees. This is comparable to the sample, as the number of employees was indicated in the questionnaire. Table 3.3 thus presents the distribution of firms according to classes of employment size for the population and for the sample in HK. One can see that most firms fall into the category of 1-5 or 6-10 employees both in the sample as well as in the population. The χ^2-test verifies that sample and population do not significantly differ ($p = 0.58$). It is assumed that the sample mirrors the population in HK according to the selection criteria.

Because sources of data were different in the PRD depending on the location of firms (see Section 3.3), a comparison of sample and population is impossible, as data about the population are not available. Data validity and quality from the PRD survey cannot be guaranteed as in the case of HK data. Because of this lack of knowledge about the data, more care must be taken with the generalisation of empirical evidence yielded by the PRD data.

Table 3.3.: Comparison sample to population according to the number of employees in electronic firms in HK (2006)

Number of Employees	Share of firms in %	
	Sample[a] ($n = 104$)	Population ($N = 4903$)
1–5	43.3	50.5
6–10	27.6	28.3
11–15	8.7	6.1
16–25	9.5	8.1
26–50	7.3	4.0
51–100	3.5	3.0
Total	100.0	100.0

[a]$\chi^2 = 3.76, p = 0.58$

Source: Calculation based on own survey conducted in DFG SPP 1233 [2007] and HONG KONG TRADE DEVELOPMENT COUNCIL RESEARCH DEPARTMENT (HKTDC) (n. d.)

3.5. SUMMARY AND DISCUSSION OF FINDINGS IN SECTION 3

E How can informality and flexibility be operationalised in customer and producer relations?

F Which methods are suitable for measuring informality and flexibility in customer producer relations?

This section has introduced and defined the GPRD as a research region with limited availability of secondary data. Additionally, the electronics industry has been identified as being the most appropriate to study informality and flexibility due to its relevance in the PRD. To operationalise informality, the five dimensions of informality (see Table 2.1) were applied. Using this method, informal interactions can be compared with formal interactions in terms of relevance, importance, frequency, etc. The three main research areas with respect to informality (recruitment, contractual arrangements, enforcement mechanisms) each study one or two dimensions of informality. It was decided to collect data about informality in an indirect way, rather than asking directly. This ensures that the same definition of informality is applied for all interactions, preventing the definition from differing between respondents. The quantitative flexibility can be operationalised using the speed involved in the different methods of interaction. The qualitative flexibility can be operationalised using the varieties of ways firms consider interacting with customers and producers. This answers research questions **E** related to how informality and flexibility can be operationalised in customer and producer relations.

In order to measure informality and flexibility, a combination of qualitative and quantitative methods was applied and assessed (research question **F**). The need for flexibility can be assessed using secondary data and literature review combined with expert interviews. The way firms deal with the need for flexibility can only be measured by collecting primary data, as there is a lack of secondary data due to the novelty of the field of research. Different research methods were applied to collect primary data on informality and flexibility in interactions with customers and producers. It was decided to follow the value chain from large HK firms via SMEs in HK to firms in the PRD, because this is the way in which flexibility is required. For the large firms in HK, a case study approach was chosen due to the diverse field of operations. For SMEs in HK, a standardised survey was conducted, but personal interviews were arranged to ensure that questionnaires were understandood. Firms in the PRD were surveyed using a standardised letter. Regional stakeholders in the PRD and HK were also interviewed. For all different measurements, the topic of informality and flexibility was touched on indirectly by assessing firms' interactions with customers and producers according to their degree of informality and flexibility. It turned out that the accessibility and the quality of interviews in HK was surprisingly high. Furthermore, the standardised, but personally conducted survey of SMEs in HK was also found to be of high quality. PRD data need to be analysed more carefully, especially with regard to possible generalisation. The following empirical sections will analyse the primary and secondary data according to their availability and suitability of the specified research topic.

4. FIRM CONDITIONS IN THE GPRD: INSTITUTIONAL CHANGE AND DEVELOPMENT OF THE ELECTRONICS INDUSTRY

This chapter focuses on the economic impact of firms in the GPRD on the global electronics industry and their competitive advantage within it. It provides empirical evidence to answer research question **G**. The chapter will:

- document how electronics firms in HK and the PRD are integrated into global value chains

- analyse the flexibility required by globally integrated firms in the GPRD

- report how the division of labour between HK and the PRD is organised and how this translates into business success

The Chinese transformation process, including the formal and informal institutional change, is described in Section 4.1. When China began to attract FDI, HK was in favour of a complete institutional setting with sufficient trading safeguards, so Western LFs started to outsource production activities there. A shortage of factory space and higher labour costs in HK led HK firms to take the opportunity of moving production activities to China during the first phase of opening, especially to the PRD. Geographical and social proximity provided HK firms with a competitive advantage over other foreign firms for success in China . Since then, China has further renewed its formal institutional environment. At the same time, the global development of the electronics industry (see Section 4.2) provides evidence for the increasing modularisation and outsourcing activities of the industry since the 1980s. A change from hierarchically organised value chains to a modularised system has taken place. In Section 4.3, the economic growth and increasing prosperity in the PRD is outlined. This results from the international integration of electronics firms into modular value chains and the simultaneous requirement for flexibility. In Section 4.4, the spatial distribution of GPRD firms integrated into global value chains is studied. HK firms tend to have a global customer base and a locally concentrated network of producers. The unique economic setting—the complete institutional environment in HK and a transitional setting in the PRD—is responsible for the distinctive organisation of customer and producer relations.

4.1. TRANSFORMATION AND INSTITUTIONAL CHANGE IN THE GPRD

The rise of the electronics industry in the GPRD is strongly connected to the opening of China and formal institutional changes (see Section 4.1.1). The attraction of FDI gave impulses for the shift of production facilities of HK firms to the Chinese mainland, especially to the Guangdong province. The similarity of informal institutions (culture, values, language, business culture) led to the decision for HK firms to move to the Guangdong province (see Section 4.1.2). The effectiveness of economic reforms in China can be proven by analysing international governance indicators (see Section 4.1.3) and economic performance figures (see Section 4.1.4).

4.1.1. Reform Policies in China and Challenges for the GPRD

China's Reform Policy

In 1978, Deng Xiaoping became the leader of the communist party in China. He started to reform the economic system in a gradual way. The other party members followed his idea of liberalising markets. China converted its economy from an administratively driven command economy to a price driven market economy, although the government still kept control over certain sectors (state capitalism). The Chinese leadership used reforms and improvements rather than replacing existing institutions. Referring to Section 2.3.1, the following reform policies were crucial for a successful transition in China: liberalisation, privatisation, industrial and foreign investment policy, enterprise income tax policy, legal reforms and financial policy (NOEL 1997). The change to the formal institutional environment aimed to achieve the following outcomes in the GPRD:

- fewer state-owned businesses and market activities

- economic growth

- increasing trade and foreign investments

- international competitiveness (ZHOU 2006)

The following paragraphs focus on different reform policies relevant for the organisation of customer producer relations of firms in the GPRD.

Liberalisation: The liberalisation of the national economy was the main focus during the first years of transition (1979–1989). The state withdrew its influence in the agricultural and industrial sector. A contract responsibility system of production in agriculture was introduced in 1979—a policy that allowed farming families to work a piece of land under contract and to keep whatever profits they earned. In most other sectors, the role of government was reduced between 1983 and 1988. Managers were given more autonomy, the government reduced emphasis on planned quotas, allowed enterprises to produce goods outside the plan for sale on the market and permitted enterprises to experiment with the use of bonuses to reward higher productivity. In essence, the central government gave away more authority to provincial and local governments and allowed them to act on their own initiatives. From the mid-1980s, the government gradually began to implement the deregulation of prices for selected goods. By the early 1990s, this had been extended to included almost all goods. Only some goods, such as rice or fuel, are still subject to price control to ensure equal distribution and to support rural households (ZHOU 2006).

Privatisation: Until the end of the 1970s, three quarters of the industrial production was generated by state-owned enterprises (SOEs). In the first phase of privatisation, SOEs' managers were given more decision-making power and incentives. After the 15th National Congress of the Communist Party of China in 1997, fundamental changes in the ownership and management of SOEs were decided upon in an attempt to address the persistent inefficiency of SOEs. Existing SOEs still cede market shares to private enterprises today. Private enterprises were set up rapidly after basic steps for private business activities were established at the beginning of 1980s (legalisation of private business activities, bank loans to private enterprises). The private sector consists of FIEs and domestic private enterprises. By now, SOEs account for 50% of the enterprises, but they are only responsible for 40% of the GIO. Private enterprises and FIE generate 55% of the GIO. The state sector relinquished its dominant role to the private sector. Private firms are allowed to select their customers and producers freely. The privatisation process started earlier in the PRD region, since it was an experimental zone for the central government. Private enterprises are of higher importance there than in the rest of China (LIEFNER 2006; HAN and PANNELL 1999; ZHOU 2006; PARK and LUO 2001:460–461).

Industrial Policy and FDI attraction: Before the reform policies were implemented, China had virtually no involvement in foreign trade. The state held a monopoly on external trade. In 1979, China began to open up with the formal establishment of foreign trade organisations, which functioned as transmitters for external trading firms, but which still followed the state-set trading plans. FDIs formed the central element of the opening and modernisation strategy in China, because they provided technology and management expertise. Because the central government decided on a gradual transformation of institutions, it was able to begin attracting FDIs, but at the same time protect its domestic firms. In 1980, the central government carried out preliminary opening experiments with a designed blueprint of economic zones in coastal regions, to which firms were attracted by the offer of special treatment. The first special economic zones (SEZs) to be opened up were SZ near HK, Zhuhai near Macau, Shantou with a good overseas Chinese network and Xiamen opposite Taiwan. These zones in the southern provinces are geographically distant from the political and economic centres of China. The initial size of zones was small, but they swelled into big cities once capital and labour flowed in and the necessary infrastructure was developed. HK and overseas Chinese firms started to move in later (at the end of 1980), capital from HK taking the biggest share. Foreign trade procedures were greatly relaxed between 1979 and 1981, which led to a higher number of customer producer relationships. SEZs are mainly characterised by:

- attracting and utilising foreign capital

- special tax incentives for foreign investments

- greater dependence on international trade activities

- export-oriented production

- sino-foreign JVs and partnerships as well as wholly foreign-owned enterprises

In 1984, the central government opened 14 coastal cities to the outside world, including GZ. They enjoyed more favourable policies than inland regions, but were not given the same status and treatment as those special economic zones.

In the beginning, Chinese law allowed foreign capital to flow in only through JVs (Chinese majority shareholding). Partners in the PRD were a necessity. The Chinese partner was responsible for providing buildings, workers and infrastructure, whereas the foreign partner provided machines and raw material. Export-oriented firms were main target group of the central government. Since the early 1990s, the central government has allowed foreign partners to become chairs of JV boards, and has authorised the establishment of wholly foreign-owned enterprises, now the preferred form of FDI. An increasing competition between foreign and domestic firms led to an adaptation of the domestic firms to new technology and prices (LIEFNER 2006).

Moreover, Deng achieved a positive result in negotiations with the British government. They agreed to return HK to China in 1997. The slogan for the reintegration of HK into China was "one country, two systems". According to this slogan, HK kept its own government, currency and laws. The reintegration of HK boosted the quantity and quality of economic links between HK and the PRD (ZHOU 2006).

Tax Policy and FDI attraction: To attract FIEs, the central government introduced preferential tax treatment (1) on imports and exports and (2) on incomes. Both are important for the organisation of customer producer relations.

(1) Soon after the opening, the Chinese government set up its processing trade policy. Processing trade refers to the business activity of importing all parts and components from abroad and re-exporting the finished products after processing or assembly by enterprises on the mainland, where the import and export of products

is subject to value added tax (VAT) rebates. FIEs carry out processing or assembly for a processing fee. To guide the flow of FDIs and encourage processing trade, the central government developed a system which divides foreign investment projects into four categories: namely encouraged, permitted, restricted and prohibited. A product catalogue was developed in 1995 (major revisions made in 1997, 2002, 2007), which listed all manufacturing goods in the different categories. FIEs which carry out projects in the encouraged category (including foreign-invested R&D centres, high-tech and export-oriented FIEs) enjoy the most preferential tax treatment, followed by FIEs carrying out projects in the permitted category (HONG KONG TRADE DEVELOPMENT COUNCIL RESEARCH DEPARTMENT (HKTDC) 2004b). Through this measure, the central government can control and guide FDI according to the attainment of national goals in different sectors and regions. This policy avoids undesired effects on the national economy and limits the power of transnational companies (HAYTER and HAN 1998). In August 2007, the categories of restricted and prohibited goods were expanded from 394 to 2,247 items in the Guangdong province (HONG KONG TRADE DEVELOPMENT COUNCIL (HKTDC) 2007d). The central government is trying to move export processing firms to the Western provinces, where they can still enjoy the tax exemption. The category of encouraged and permitted goods was reduced to capital-, knowledge- and technology-intensive products. An upgrade of technology was initiated in the Guangdong province. Firms in Guangdong can continue to engage in processing trade of products under the restricted category, provided that they pay customs duty deposits. These new regulations hit electronics firms in Guangdong especially hard, as many electronic items are now in the restricted category. Since the beginning of the Chinese opening policy, processing trade has solved China's shortage of capital and technology-related expertise. Moreover, it has functioned as a job-creating machine. 70,000 of the 90,000 export processing firms were located in Guangdong in 2006. More than 16 million people are estimated to be working for processing trade enterprises in Guangdong. 47% of the total foreign trade results from export processing enterprises. The policy forces firms in the GPRD to make major changes—either to upgrade or to relocate (see paragraph below). Due to the sudden changes to the categories of restricted and prohibited goods, it is expected that firms in the GPRD will close down, because they are not able to relocate or upgrade immediately (THE GREATER PEARL RIVER DELTA BUSINESS COUNCIL 2007). The recent economic crisis in 2008/2009 has accelerated the closing down of companies. In autum 2008, 15% of firms in the PRD had to close their businesses.

(2) At the beginning of China's opening, the central government reduced the income tax for FIEs in order to attract FDIs. FIEs established in special economic zones, hi-tech industrial zones and state-level economic development zones had to pay enterprise income tax at a rate of 15%. JVs were granted to FIEs. FIEs in costal open areas and provincial capitals were subject to a tax rate of 24%. SOEs had

to pay 55% income tax and domestic enterprises 33% (HONG KONG TRADE DE-VELOPMENT COUNCIL RESEARCH DEPARTMENT (HKTDC) 2004b). Effective from January 2008, the central government changed the enterprise income tax law. It unifies the income tax systems to 25% for both domestically-funded enterprises and FIEs. Most of the enterprises with foreign investment had previously enjoyed a lower tax rate. This was announced only eight months before taking effect. Grand-father clauses are included in the new law to ensure a smooth changeover. Exemp-tions of 15% income tax only apply to high-tech firms. It is still undecided whether special treatment should continue for firms in SEZs. Besides the new financial bur-den of tax on imports and exports, FIEs in the Guangdong province also suffer from higher income tax rates (ETGEN 2007; ZHOU 2006).

Legal Reforms: The central government in China put into effect four major changes in order to reform the legal system. (1) The central government changed the prop-erty rights from emphasising socialist public ownership to protecting private own-ership. Personal and public property rights were declared protected by law, thus reducing the risks of doing business. This process has still not been completed suf-ficiently, particularly the protection of IPRs. (2) Lawyers are allowed to practice again in China. Great efforts have been made to improve the quantity and quality of lawyers with many training and exchange programmes. The quality and education of judges is improving, but is still not sufficient for FIEs to ensure a fair judgment. This affects how firms choose to settle disputes with customers and producers. (3) The introduction of more business-related laws has led to an improvement of the business environment and legal protection of business practice. It was also partly carried out in compliance with the World Trade Organisation (WTO) rules. Rele-vant laws include the enterprise law and banking law. (4) The establishment of new laws has seen major improvements, but enforcement is still lacking. To make the "rule of law" applicable in economic and social life, the government began to put emphasis on legal procedures. More parties involved in business stand to profit from cases heard and disputes resolved. Although new laws are building market-related institutions in China, their enforcement needs major improvements, as corruption of Chinese officials is still a daily occurrence. Moreover, China does not offer in-ternational security standards for protecting international contract conclusion and enforcement. In case of conflicts between international parties, arbitration is a com-mon way of settling disputes. While HK possesses an excellent dispute settlement system, China lacks the required institutions (ZHOU 2006). For example, the UN model law on *International Commercial Arbitration (1985)* has not yet been im-plemented, but China's arbitration law is under revision for improvement (UNITED NATIONS COMMISSION ON INTERNATIONAL TRADE LAW (UNCITRAL) n. d.). Even if the *Convention on the Recognition and Enforcement of Foreign Arbitral Awards (1958)* is ratified in China, it is nevertheless difficult and time-consuming to deal with the Chinese courts in order to enforce a foreign arbitration award reached with international customers and producers.

Financial Policy: The implications of the economic reforms for the financial sector are tremendous. The privatisation of enterprises and the liberalisation of prices require (1) a bank reform to ensure business financing and (2) the introduction of a new functional tax system to generate public revenue. Although the central government has reformed the tax law, the enforcement of a standardised tax system is difficult, because firms are used to tax exemptions. Furthermore, the central government has introduced a two-tiered banking system. When FIEs seek financing in China, Chinese commercial banks are allowed to accept guarantees from foreign shareholders, but the central government still controls the establishment of foreign banks and branches. In 1995, China linked the RMB to the USD for currency convertibility. In the course of its accession to the WTO, China had to open its banking sector. Since 2007, China has opened up for foreign banks but has limited the establishment of branches. Chinese banks are still not subject to international competition. Chinese banks do not provide a good alternative to foreign banks, especially for FIEs. This limited availability of capital for FIEs reduces the attractiveness of China and increases business risks. Therefore, many firms tend to organise their capital flows to customers and producers via HK (LIEFNER 2006; OKURA 1996; AGARWAL 1999; HONG KONG TRADE DEVELOPMENT COUNCIL RESEARCH DEPARTMENT (HKTDC) 2004b).

Newest Trends and Challenges for Firms in the GPRD

Upgrading or Relocating: Since the beginning of 2000, the central government has started to encourage firms to relocate to the central and western regions. Projects concerning energy and transportation infrastructure are especially likely to enjoy preferential treatment. As described above, the Chinese government attracts and controls FDIs in western provinces through tax exemptions for imports and exports for certain goods and also by offering a reduced rate of income tax (15%). Just as they make the western provinces attractive for export-oriented FIEs, they reduce the attractiveness of the Guangdong province for those enterprises. In 2007, the central government expanded the list of restricted and prohibited goods for tax-free import and export for enterprises set up in Guangdong (HONG KONG TRADE DEVELOPMENT COUNCIL (HKTDC) 2006b). China continues to support the Guangdong province in terms of FDI attraction for capital- and technology-intensive industries through lower income taxes, despite the fact that FIEs actually should have adapted to the new enterprise tax law in 2008. Moreover, environmental production standards were introduced. Enterprises failing to meet eco-standards will be refused registration in Guangdong (HONG KONG TRADE DEVELOPMENT COUNCIL (HKTDC) 2007a). These three changes to the legal systems hit Guangdong province especially hard, as many FIEs relying on export processing are set up there. They must either decide to adopt a new upgrading strategy to meet the defined requirements of capital- or technology-intensive products, or they must relocate to the western provinces and continue business operations there (BERGER and LESTER 1997; HONG KONG TRADE DEVELOPMENT COUNCIL (HKTDC) 2006a). This greatly affects customer producer relations. Firms must not only upgrade their own business activities, but those of their producers as well. When firms relocate, they either have to take their producers with them to the new location, or they must build a new producer network. Moreover, production costs in the PRD have risen due to increased labour costs (minimum wages and social security costs) and additional utility charges (such as costs for electricity). By July 2007, the RMB exchange rate had crept up to 9.3% against the USD. While most HK manufacturers still receive USD payments for their exports, the appreciation of the RMB means that the part of production costs settled in RMB would increase in terms of the USD, which diminishes export margins (HONG KONG TRADE DEVELOPMENT COUNCIL (HKTDC) 2007c). Moreover, in January 2008, the new Labour Contract Law came into effect, which protects employees' rights even further. The new law brings new requirements for the termination of labour contracts. It encourages employers to sign a long-term labour contract with their employee. This means a loss of manufacturing flexibility for the firms in the GPRD.

The Opening of China for FDI in Services: HK, like most Western countries, has been a member of the WTO since 1995. China, however, did not enter the WTO until 2001. After that, China had to open its economy further and adapt its governmental functions to a more liberalised policy. Numerous government regulations and internal rules had—and still have—to be revised. Government efficiency and effectiveness in implementing laws is also in question, judging by international standards. In other words, with the WTO accession, the business environment in China had to improve rapidly to avoid penalties and retaliation. China was requested to offer a standardised and transparent legal environment for FIEs, a better protection of intellectual property rights (IPRs) and a reduction of tariffs and non-tariff trade barriers. However, after just seven years of WTO membership, China's reforms on theses issues are obviously not yet sufficient. Moreover, service industries had to be opened for foreign investments within five years of accession, including telecommunication, banking, insurance, securities, foreign trade and legal services. For the Guangdong province, the opening of the banking and insurance sector is of particular relevance. Banking and insurance firms could serve FIEs with branches in their neighbourhood, making communication and administrative procedures easier. In 2006, China's WTO five-year-implementation phase for opening its banking and insurance sector for foreign firms ended. But it must be noted that some regulations for implementations have still not been finally set (SCHUELLER 2002; YEUNG 2002; HONG KONG TRADE DEVELOPMENT COUNCIL RESEARCH DE-

PARTMENT (HKTDC) 2000a; HONG KONG TRADE DEVELOPMENT COUNCIL RESEARCH DEPARTMENT (HKTDC) 2001).

CEPA 2004–2007 and other Trade Agreements: To foster the cooperation between the Guangdong province and HK, the Closer Economic Partnership Arrangement (CEPA) was established in 2004. CEPA derives essentially from a response by the HK business community to China's WTO accession. It is based on the idea of extending free trade between HK and the Chinese mainland in order to gain significant benefits to offset the impact of post-WTO entry competition. Gradually, HK firms were granted more preferential treatment than other FIEs in Guangdong. CEPA came into force in five steps from 2004 to 2007. Zero tariffs were applied on goods originally from HK (tariff rates which would otherwise apply range from 4 to 30%). The corresponding catalogue contains 273 items. Under CEPA, 90% of the HK domestic exports to the mainland are free of tariffs. Special emphasis is placed on the liberalisation of different service sectors, for example management consultant services, exhibitions and conventions, real estate, distribution services, logistics services, legal services, banking, insurance (HONG KONG TRADE DEVELOPMENT COUNCIL RESEARCH DEPARTMENT (HKTDC) 2003a; HONG KONG TRADE DEVELOPMENT COUNCIL RESEARCH DEPARTMENT (HKTDC) 2004a; HONG KONG TRADE DEVELOPMENT COUNCIL (HKTDC) 2007b; HONG KONG GENERAL CHAMBER OF COMMERCE (HKGCC) 2003). The mainland had opened service industries and markets to HK businesses prior to the overall opening to international businesses required by WTO rules. To strengthen international trade, China concluded several bilateral free trade agreements and reached a regional free trade agreement with the Association of Southeast Asian Nations (ASEAN). From 2007 to 2010, trade in goods and services with the ASEAN members will be liberalised.

Economic Crisis 2008: As a financial centre, HK has been hit heavily by the recent financial crisis. To help SMEs in HK get credit, a Loan Guarantee Scheme and a SME Export Marketing Fund were implemented. Moreover, the PRD as an export-oriented region suffered from the sharp decline in demand from leading global firms. The customer producer relationships were put under pressure as orders were reduced and competition grew. The HK government is trying to strengthen cooperation with Guangdong in order to maintain the competitive edge of the GPRD. Mainland enterprises were encouraged to use HK as a platform to develop international business. The central government in China will further adjust the tax rebate rates for exports, improve the labour system to reduce the burden on enterprises, facilitate the development of the domestic market for processing trade enterprises and develop a guarantee mechanism for SME financing. Despite these efforts, about 15% of firms in the PRD have already closed down (INVEST IN HONG KONG 2008).

4.1.2. Shaping of the Business Environment by Informal Institutions

The business environment in HK and the PRD is not only shaped by formal institutions, but is also characterised by the Chinese way of doing business while relying on informal institutions. The way the Chinese acquire and organise their business differs from Western ways. HK firms can take advantage of their special ability to deal simultaneously with both the Chinese transitional setting and global markets (LI et al. 2004:71). They are famous for their Janus-faced business organisation (MEYER et al. 2009). The fact that the Chinese way of doing business has not been sufficiently studied and considered by foreign firms causes them to fail in China. Chinese firms are on their way to imitating HK firms in dealing with both settings.

Corresponding to Section 2.3.5 in the conceptual part of this work, informal institutions must be taken into account when analysing the environmental conditions in the GPRD. Chinese business organisation relies on the personal and business networks of individual merchants, which results in informal rules difficult to understand for Westerners. This business model, best represented by the HK experience, is characterised by its flexibility and the absence of state patronage (LOH 2002:14). The following traits describe the Chinese way of understanding and organising business. They give an insight into what rules informally shape the business environment.

Family Business: Small and medium sized family businesses are the most common type of firms. The reliance on ethnic and community bonds in business operation is widespread. Firms organise themselves into tight mutual aid associations. The authority of top management remains in the hands of family members with an authoritarian leadership style. Decisions can be made very quickly. Family businesses fit very well with the Chinese culture as the family is the centre of society. The cohesion of the family and the firm provides stability to the business and builds up reputation, especially during the initial years of foundation. Younger generations in HK have obtained a more extensive education in the West and have gained deeper understanding of the way in which Western businesses work. This has provided them with a combination of the Chinese and the Western way of thinking, which increases their scope for flexible firm organisation. Today, family businesses, especially in the PRD, face three problems. Firstly, the process of transition is opening a gap between family-controlled businesses on the one hand and open market competition and the successful attraction of outside talent on the other hand. Secondly, there are problems in handling the growing size of their businesses. A larger firm requires more management skills, which are not always available to Chinese family firms. Thirdly, it is a big challenge to combine growth and the necessary professionalisation of management with the typical flexibility and reaction rate of the firms (LOH 2002; ZHOU 2006:45-67). Nevertheless, family-owned firms are not just a hallmark of HK and the PRD - their flexibility makes them the crucial factor in the region's success.

Business Networks: In recent years, several studies have been carried out claiming to investigate the "Chinese economic model". An important finding of these studies is that behind the aimless and opaque appearance, there is a complex and competitive organisational model (LOH 2002). Network structures among family businesses are fundamental to these flexible firm organisations. In Chinese society, the sense of community and responsibility plays an important role. This also extends into the business world. Therefore, strategically important positions in a firm are occupied by family members, and firms belonging to individual family members work together closely. Firms are often supported by family members and family members are frequently given preferential treatment in commissioning. Swift and direct communication is possible, even concerning large assignments, due to a high degree of trust. Additionally, the success of the Chinese way of doing business can be attributed to an enlarged contact network among classmates, military service colleagues, home town friends and friends of friends (guanxi). Trust and partial lifelong reciprocity between people who have guanxi with each other results in help when difficulties arise, access to rare resources and the opportunity to obtain political support for personal projects. Loyalty and respect towards the leaders and towards business partners is responsible for stability and smooth business operations. Executives need to spend much of their time dealing with partners and exercise extreme control over strategies. These informal interactions are essential to a company's high degree of flexibility. The main reason for a corporation's success is the degree to which it is embedded in guanxi-networks (LOH 2002; PARK and LUO 2001). The majority of the literature today accepts the prevalence of guanxi as a Chinese asset of personalised social capital (LI 2007). Recently, functional or utilitarian aspects of guanxi have become more important than emotional aspects (PARK and LUO 2001:457). Guanxi has transformed into an informal channel of doing business, which promises to be more successful than relying on formal methods alone. This finding goes beyond the standard expectation of institutional economics, which maintains that informality in China is a transitional phenomenon and a substitute for formal institutional support (XIN and PEARCE 1996).

Experience in Risk-Taking: Many studies emphasise not only network structures as a crucial factor of success, but also the company's ability to compete with others and grow despite market uncertainty. Over and above that, many business families who have been successful in HK for decades had already been successful businessmen in China before fleeing to HK during the Cultural Revolution (LEE 1997:195-205). However, unpredictable political and economic circumstances both during and after the British rule forced HK companies to develop strategies and measures for reducing insecurity on a permanent basis. HK firms tend to rely on intuition to anticipate eventualities, rather than on quantitative information and calculations.

Other typical methods include spreading the risk over several persons and companies, building up financial reserves, diversifying business fields and selecting personnel carefully (SULL 2005). This pool of experiences leads to the conclusion
that insecurity not only holds risks, but also offers opportunities (LOH 2002; ZHOU
2006:45-67).

Guerrilla Entrepreneurship: HK companies have learned how to manage insecurity and how to recognise new, short-term business opportunities successfully using
small family businesses working within a large network. This "guerrilla strategy"
facilitates their international competitiveness (YU 2000). For a very short time, the
companies specialise in goods with a high profit margin and exploit the situation by
overstocking the world market. Established companies can only act with a delay in
such cases. After short-term profits, HK companies leave the market and focus on
other products. The electronics industry in particular is very well known for this
strategy. The Tamagochi, an electronic toy invented in Japan, was mostly produced
in the PRD after its successful introduction to the market. Japanese companies were
not able to modify their production quickly enough, so PRD companies drew high
profits. A widely spread and reliable supplier network is a big part of the companies' guerrilla strategy (LEE 1997:195-205). After accepting a production order, a
HK company not only has access to its own factory but also to those of their partners and suppliers. If their own capacities are exhausted, orders can be passed on.
Hence, there is no need to decline additional orders. Because of the great diversity
of supply networks, low investment is required, fluctuations of demand can be bal-
anced out much more effectively and the whole process accelerates thanks to the
geographical proximity of all participants. This allows the transaction costs to remain low. Emerging Chinese firms try to follow HK firms' most effective practice
model (LOH 2002; ZHOU 2006:45-67).

4.1.3. Putting China's Institutional Environment into a Global Perspective

The central government is gradually reforming the rules of law in China. The reform process is combined with the informal institutional environment, which is only
adapting slowly to the new rules of law. The annual collected factors of the *economic freedom index* calculated by the HERITAGE FOUNDATION (n. d.) provides a
longitudinal and a cross-regional analysis to put China's institutional environment
into a global perspective. The economic freedom index can be taken as an objective
measurement with which to follow the development of China's institutional environment. The economic freedom factors cover the firm-specific environment in a
country. It consists of ten factors. Each one of the ten factors is graded using a scale
of 0 to 100, where 100 represents the maximum economic freedom. In evaluating
the criteria for each factor, the HERITAGE FOUNDATION (n. d.) has used a range
of authoritative sources. The first factor, for example, *business freedom*, consists of

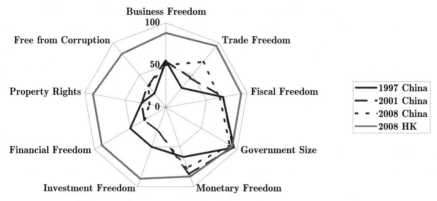

Figure 4.1.: Development of the institutional environment in China
Source: HERITAGE FOUNDATION (n. d.)

indicators collected by the WORLDBANK's "Doing Business" study. It covers the
ability to create, operate, and close an enterprise quickly and easily. Cumbersome,
redundant regulatory rules are the most harmful barriers to business freedom. *Trade
freedom* is a composite measure of the absence of tariff and non-tariff barriers that
affect imports and exports of goods and services. Sources for that are World Bank
indicators, WTO indicators and Economist Intelligence Unit measurements. *Fiscal
freedom* is a measure of the tax burden from the revenue side. *Government size* is
defined to include all government expenditures. Ideally, the state will provide only
true public goods, with an absolute minimum of expenditure. *Monetary freedom*
combines a measure of price stability with an assessment of price controls. *Investment freedom* is an assessment of the free flow of capital, especially foreign capital.
Financial freedom is a measure of banking security as well as independence from
government control. State ownership of banks and other financial institutions, such
as insurer and capital markets, is a burden on efficiency, and political favoritism has
no place in a free capital market. *Property rights* is an assessment of the ability
of individuals to accumulate private property, secured by clear laws that are fully
enforced by the state. *Freedom from corruption* is based on quantitative data that
assess the perception of corruption in the business environment, including levels of
governmental, legal, judicial, and administrative corruption. This factor reverts to
the corruption perceptions index of TRANSPARENCY INTERNATIONAL (n. d.). The
last indicator, *labour freedom*, is not taken into account here because of missing
data for the years in question. Because the economic freedom index summarises a
range of other indicators, it seems to be the most suitable index with which firstly
to follow China's development over a period of years, and secondly to compare
China's state-of-the-art institutional environment to that of HK.

Figure 4.1 shows China's development in terms of the nine factors comparing the years 1997 (HK's return to China), 2001 (WTO accession) and 2008 (today). To put China's figures into perspective, HK's data are added for 2008. When comparing the China curve and the HK curve, one can see that HK reaches between 80 and 90 points out of 100 for all factors of the economic freedom index. HK provides excellent institutions to support firms and enables free flow of investment and trade. China is still struggling to reach half of the points for some factors.

When examining the development of China over time in terms of the different factors, one notices that the business freedom factor has not increased within the last eleven years. The number of days and procedures for starting and closing a business has not been reduced. Moreover, the indices for getting credits and protecting investors have hardly changed (WORLDBANK n. d.). In contrast, there is a large improvement in the area of trade freedom. The scale has increased from 30 to 70 points. China's accession to the WTO, the CEPA agreement and the profits made by exports have boosted the opening of China for foreign trade. Fiscal freedom in terms of tax burden has remained on the same high level for the last eleven years. It is expected that the fiscal freedom will improve within the coming years due to the new enterprise tax law reducing the rate for Chinese firms. In terms of government size, China has reached the HK level and does not need to hide its competence. In terms of monetary freedom, China has improved greatly in recent years. Prices have become more stable, which encourages investments. However, when comparing 1997 to 2008, one notices that China has become less attractive in terms of the free flow of capital (investment freedom) and the high dependency of the banking system on the central government (financial freedom). In terms of services, China has still not opened enough for foreign firms to protect their own businesses sufficiently. Even if the WTO accession demands this opening, it is only taking place very slowly. These results confirm what has been said about the limited opening for services in Section 4.1.1. The protection of property rights is also in a poor condition and is always criticised by firms. The scale has even fallen from 30 points in 1997 to 20 points in 2008. The corruption factor also shows China to be in a bad position. Although improvements could be seen in the increase from 22 to 33 points, the overall level is low. China's major weaknesses in terms of institutions are its susceptibility to corruption, its poor protection of property rights, especially IPRs, and its limited opening for financial services. Although those weaknesses are well known, there has hardly been any improvement in recent years. China could have taken a major leap forward in terms of free trade flow. Taking objective measurements into account shows that China may have improved in some areas in recent years, but is still in a poor condition in others. In comparison with HK, China can only compete in terms of government size (government expenditures) and monetary price stability. In all other fields, HK provides firms with much better and more stable conditions. When comparing China as a whole with the PRD, one can expect that trade freedom for HK firms is higher in the PRD than in the rest of China due to the CEPA. Additionally, the financial and monetary freedom is less relevant in the PRD, because firms can also get loans from HK banks. Government size and control is generally lower in the PRD than in the rest of China. It can be expected that the PRD is sightly better situated in terms of the institutional environment than China as a whole. The next section focuses on the development of economic figures in the GPRD to complete the picture of the situation in the research region.

4.1.4. The GPRD in Figures

The outsourcing strategy of Western LFs in the 1980s was closely linked to the rise of the four tiger states: HK, Taiwan, Singapore and South Korea. They bene-fited mainly from the transfer of production capacity and knowledge from Western firms fostered by strong political support. The rise of HK was closely connected with the growth of the PRD region due to two parallel developments. On the one hand, growing prosperity in the 1980s in Europe and the United States (US) led to an increasing demand for consumer products, especially electronic devices (see Section 4.2). As competition between Western firms grew, they looked for attrac-tive production locations with cost-efficient assets. At the same time, HK became an attractive location for production outsourcing for leading Western firms because of its well functional legal system. However, factory space became limited in HK due to the increasing foreign demand for the production activities of HK firms. In 1978, Deng Xiaoping became the leader of the communist party in China. Under his leadership, a SEZ was established in SZ in 1980 because of its proximity to the economically successful city of HK (see Section 4.1.1). Deng hoped that China would profit from HK's economic growth with this experiment, and in case of fail-ure, it was politically far removed from Beijing. But the SEZ in SZ did not fail. Deng allowed foreign goods and capital to be moved into China. Investments from HK firms were especially welcome. By then, HK firms were following Deng's in-vitation to set up plants in SZ. The proximity to HK and the opportunity of keeping control of production activities in the PRD led HK entrepreneurs to take the risk of producing in China. The success of HK firms led Taiwanese and other foreign firms to follow this path. Further liberalisation policies, China's accession to the WTO and the introduction of the CEPA increased the attractiveness of the entire PRD.

The results of the tremendous change of institutions—the economic growth in the PRD—can be verified by the increase of FDIs, GIO and the trade volume. Ta-ble 4.1 shows the time series (1980–2005) of economic figures in Guangdong, GZ and DG. Since 1980, a rapid increase of FDIs can be recognised, with the largest boom coming between 1990 and 2000. Since 2000, the FDI flow has stagnated in both the Guangdong province and in GZ (compound annual growth rate (CAGR) 2000–2005: 0% or −2% respectively). In DG, the rapid flow has continued. GZ and DG were responsible for nearly half of all FDIs in the Guangdong province in

Table 4.1.: Time series analysis of economic indicators in the PRD 1980–2005

	1980	1985	1990	1995	2000	2005	CAGR 80–05	CAGR 90–05	CAGR 00–05
FDI inflow (in 100 mill. USD)									
GD	1.2	5.2	14.6	101.8	122.4	123.6	20%	15%	0%
GZ	0.1	1.0	1.9	21.4	29.9	26.5	24%	19%	−2%
DG	0.0	0.1	1.0	6.7	10.9	28.2	53%	25%	21%
GIO[a] (in 100 mill. USD)									
GD	165.8	182.1	397.6	1,164.0	2,042.0	5,085.8	15%	19%	20%
GZ	58.7	60.6	92.5	206.3	374.5	826.2	11%	16%	17%
DG	3.9	5.9	20.6	62.6	183.6	545.7	22%	24%	24%
Trade Volume (in 100 mill. USD)									
GD			419.0	1,039.7	1,701.1	4,280.0		17%	20%
GZ			41.8	167.0	233.5	534.8		19%	18%
DG			10.8	153.9	320.5	743.7		33%	18%

[a]since 1995 new classification decreasing values slightly

Source: Own calculations based on GUANGDONG PROVINCIAL BUREAU OF STATISTICS (2007), GUANGZHOU MUNCIPAL STATISTICS BUREAU (GMSB) (2007), DONGGUAN MUNICIPAL STATISTICS BUREAU (2007)

2005. GZ is characterised by more local Chinese enterprises, whereas DG remains a well known destination for Taiwan and HK investments. The different characteristics can also be seen in the development of the GIO. Despite a rapid growth of GIO in both cities, DG was able to reach a CAGR of 22% from 1980–2005, whereas GZ only reached 11%. This is clearly a consequence of more FDIs in DG during that time. Foreign firms can usually reach a higher GIO. Despite this fact, the GIO in GZ has taken the edge off DG in recent years. Looking at the CAGR 2000–2005, DG still registered a higher growth rate of GIO than GZ, but the gap between both was reduced. This shows that GZ became more economically powerful, despite the fact that GZ hosts more domestic firms. Besides a tremendeous increase in trade, DG, with its emphasis on foreign firms, has higher foreign trade figures than GZ. Although GZ firms have a higher GIO, they focus more on the domestic market, whereas DG firms are oriented towards global markets.

The last figure shows how tremendous the economic growth in the PRD has been. The following figures from 2006 show how important the PRD is for China, which role firms in GZ and DG play in the PRD and what impact the electronics industry has for the economic development of the region:

Impact of PRD electronics firms for China: The PRD is responsible for 30% of
Chinese exports, 18% of FDIs and 12% of the GIO. The impact of the electronics industry in the PRD for China is even higher. Electronics firms in the
PRD account for 38% of all Chinese electronics exports and they are responsible for 32% of the GIO reached by the Chinese electronics firms.

Impact of GZ and DG electronics firms for the PRD: Electronics firms in DG
generate 13% of the GIO of all electronics firms in the PRD. Electronics
firms in GZ account for 7.2% of the GIO in the PRD and 6.9% of the exports
of all electronics products in the PRD.

Impact of HK firms in the PRD: 32% of China's FDI inflows originally come
from HK, Taiwan or Macau. In the Guangdong province, these numbers are
even higher because of the proximity to HK. In Guangdong, 47% of all FDIs
come from HK, Taiwan or Macau, in GZ 43% and in DG as much as 58%. It
is assumed that not all FDI come direcly from HK firms, but also originate in
other countries. They are only invested via HK. How high the share is cannot
be determined from the official statistics. This also results in a high share
of GIO generated by FIEs from HK, Taiwan or Macau. In DG, firms from
HK, Taiwan or Macau are responsible for 49% of the GIO, while in GZ they
account for 26% and in Guangdong province for 30%.

GDP per Capita: As a result of the economic boom described above, the GDP per
capita has grown tremendously. Calculations for 2006 are difficult, because
the official figure is based on the registered population. Migrant workers are
not included. For 2005, figures are available based on the census population.
For the Guangdong province, the GDP per capita was 2,882 USD. In the PRD,
this figure rose to 4,865. In DG, GDP per capita was 6,142 in 2005 and in GZ
6,574. The slightly higher figure in GZ can be explained by its functionality
as a service and financial centre in Guangdong (ENRIGHT et al. 2006).

As previously indicated, growth differences between the cities within the PRD can
be recognised. A shift-share analysis was conducted to characterise the growth differences between cities and to establish which role is taken by the research cities
DG and GZ (SCHAETZL 2000). The differences in economic growth are measured
by the Total Net Shift (TNS). The TNS is the difference between the actual change
of GIO in a city over a certain time period and the potential change, assuming that
the economic activity of one city would have developed in accordance with the average for the entire PRD region. This deviation is measured in percentage points. A
positive value corresponds to the above average growth of a city in comparison to
the entire PRD region, and a negative value to below average growth. Reasons for
that can be quantified by the Net Differential Shift (NDS), which relies on an initial

locational effect, and the Net Proportionality Shift (NPS), which is based on an effect of the sectoral structure. A positive (or negative) locational effect (NDS) means a relative locational advantage (or disadvantage), because growth of certain sectors is encouraged (or limited) in a city due to other locational characteristics (for example poor infrastructure or accessibility, certain city government regulations). The structural effect (NPS) of a city is positive (or negative) when the share of sectors with above average growth is higher (or lower) than for the entire PRD due to a beneficial split of industries. For the PRD region, the economic growth is measured by the change of the GIO in the five major industrial sectors (Electronics, Petroleum and Chemistry, Textile and Garments, Food and Beverage, Building Materials).

Comparing the results of the shift share analysis of the GIO in the PRD from the periods 1990–2000 and 2000–2006, the characteristics of different cities can be shown (see Table 4.2). In the period 1990–2000, cities in proximity to HK showed an above average development in comparison to the total development in the PRD region. DG (54%), SZ (16%) and Huizhou (54%) had positive values in TNS due to the close proximity to HK, while Zhuhai (19%) and Foshan (25%) are in spatial proximity to Macau. The development in the first few years was a matter of proximity to HK, which can be confirmed by the high positive locational effects in those cities. Accessability caused by the development of infrastructure was the basic argument. GZ indicates a negative value. Although the provincial capital, Guangzhou, was already well linked to HK, the accessibility of the remote areas of the province was poor. The sectoral structure had only a small impact on the economic growth in this period. This picture changed when looking at the development of the GIO between 2000 and 2006. DG and SZ still benefited most in the PRD region, which can mainly be explained by the location. For other cities, the structural effect drove development more than the location effect (for example Zhuhai and Huizhou). Although GZ still had a below average development, the city profited more from its sectoral structure than from its location. In general, a shift from advantages in location to advantages in sectoral structure is recognisable in the PRD. This is a consequence of the high investment in infrastructure. For future development, it can be expected that the remote cities, especially in the west, will report the highest growth rates (GREATER PEARL RIVER DELTA COUNCIL 2006). HK firms view the progress of the western PRD positively, as they are looking for expansion within the PRD area rather than moving to other provinces, as desired by the central government.

A rapid increase in FDI, GIO and trade has occurred since China opened the region for FIEs. The geographical and cultural proximity to HK has contributed greatly to its economic growth. HK firms have invested heavily in the PRD and have provided expertise in technology and management for the Chinese firms. Those investments were driven by (1) the introduction of market-like institutions in China (see Section 4.1.1 and 4.1.3), (2) the close links between HK and China shaped by

Table 4.2.: Shift-Share analysis of the gross industrial output in the PRD

	1990–2000			2000–2006		
	TNS[a]	NDS[b]	NPS[c]	TNS	NDS	NPS
	Total	Location	Sectors	Total	Location	Sectors
Dongguan	54%	56%	−1%	40%	32%	8%
Shenzhen	16%	6%	9%	37%	21%	16%
Zhuhai	19%	26%	−6%	1%	−17%	18%
Huizhou	54%	49%	4%	−15%	−46%	31%
Foshan	25%	21%	4%	−70%	−24%	−46%
Zhongshan	−60%	−65%	5%	38%	42%	−4%
Guangzhou	−48%	−54%	−5%	−29%	−23%	−6%
Jiangmen	−63%	−37%	−26%	−87%	−58%	−28%
Zhaoqing	−96%	−84%	−12%	−174%	−146%	−28%

[a]Total Net Shift
[b]Net Differential Shift
[c]Net Proportionality Shift

Source: Own calculations based on GUANGDONG PROVINCIAL BUREAU OF
         STATISTICS (2007)

common informal institutions (see Section 4.1.2) and (3) the increasing global demand for products manufactured in the PRD, especially electronics products (see
Section 4.2). Although the PRD has reported amazing economic growth rates, differences in growth between the PRD cities were discovered. This section has only
given a brief overview of the development in the PRD. As the emphasis of this
work is on the electronics industry, the following section stresses the development
and characteristics of the electronics industry in the GPRD and its integration into
the world economy.

4.2. GLOBAL DEVELOPMENT OF THE ELECTRONICS INDUSTRY

This section gives an introduction to how the electronics industry is globally organised. It documents major steps in its development since the 1960s. It describes the
shift from integrated firms to modular value chains, concentrating on the participation of firms in developing countries.

The development of the global electronics industry can be divided into four areas: growth of demand, modularisation, shorter product life cycles and outsourcing.
The development phases of the electronics industry and the emergence of modular

value chains are the results of changes which began in the US. For most of the twentieth century, the electronics industry in the US was dominated by large, vertically integrated firms with component divisions. In the 1960s and 1970s, with the push for better semiconductors for military use and aerospace application, the growth of the electronics industry accelerated. In the 1980s, the consumer electronics industry began to develop rapidly with the introduction of personal computers. *Demand* for consumer electronics products grew tremendously.

A phenomenon developed which still poses challenges for the electronics industry today—speed-to-market necessitated by *shorter product life cycles*. BAYUS (1998) and KAIPIA et al. (2006) studied the reduction of product life cycles. BAYUS (1998) investigated the mean length of product life cycles for products made by internationally renowned electronics firms. Life cycles range from 2.45 years (Samsung) to 5.47 years (IBM). KAIPIA et al. (2006) complemented his research by studying an augmentation of product families and product introductions per year, which is not only a sign of increasing product diversity, but also of a shorter life time of product series. They found that production and delivery quantity must be planned on a weekly basis, which shows the high volatility of markets. Moreover, they discovered huge differences in the planning accuracy of firms at different stages of the value chain. Planning accuracy of CMs is relatively poor. Even for the following week, inaccuracy of plans comes to more than 70%. CSs have a good planning accuracy für standardised material, because raw material planning requires long planning horizons, but planning accuracy is worse for variable materials. Their results show the high degree of flexibility expected from electronics firms at the lower end of the value chain.

The increasing market pressure of quick response was no longer manageable for US firms in the age of the new economy. US firms changed their production structure from the earlier in-house production to wholly-owned component divisions. Thus began the *modularisation* process in the electronics value chain.

At the same time, the Japanese electronics industry began to develop. Instead of relying on component division of production, the Japanese built a network model of captive suppliers in Japan. The practice of Japanese firms seemed to work more successfully, with US firms losing market share. Therefore, in the 1990s the US LFs started to change from in-house manufacturing and component divisions to external suppliers in Asia, Europe and North America. The era of production *outsourcing* started. But whereas Japanese firms used a tight supplier network to meet the demand for flexibility, Western firms—especially those from the US—tended to outsource their entire production capacities to large CMs. This enabled them to compete with the Japanese firms in cost and production organisation. Hewlett Packard and IBM led the way, selling most of their worldwide manufacturing facilities to CMs (see Section 2.2.1) such as Solectron and Flextronics (STURGEON

2003:12–13). Standardised protocols were developed to hand off design files to CMs and manufacturing processes became highly automated.

With the internet boom in the mid-1990s, new US-firms (for example Sun Microsystems) appeared which never established their own manufacturing sites. Some CMs emerged in North America, but later the assembly and even some of the designing of notebooks and desktop personal computers was outsourced to "original equipment" and "original design" CMs based in Taiwan, Singapore and HK. By the end of the 1990s, much of the manufacturing capacity of the Taiwan-based and HK-based CMs had shifted to mainland China, especially to the Guangdong province. Big US CMs established a network of factories on a global scale. Around that time, the European communication companies (Nokia, Ericsson, Alcatel) also started to outsource their production. By the end of the 1990s, value chain "modules", such as design or production, were highly developed with a global scope. Even though the internet boom began to stall in 1999, large outsourcing deals continued to be announced. The first Chinese CMs appeared during that time, and CMs began taking over more functions: distribution, product development and the development of company brands. The value chain was no longer organised in a strong hierarchical form controlled by US firms, but had developed into a modular value chain (STURGEON 2006b). Even Japanese firms have recently established large production alliances for technology development, patent sharing and joint component divisions. Others have started to relocate low-end in-house operations to low-cost locations such as China. At the beginning of the 2000s, large Japanese electronics firms, for example NEC and Sony, started to sell their production facilities to CMs. By then, Japanese firms had also started to modularise their production (STURGEON 2006b).

It could be shown that the electronics industry has been driven by a high dynamic since the 1960s. With increasing prosperity during the economic boom of the 1970s and 1980s, the demand for electronics products grew. This went along with the phenomenon of shorter product life cycles, which arose from the demand of a greater speed-to-market. To better adapt to this requirement, electronics firms in the US, and later in Japan, started to modularise and outsource production activities to CMs in Asia. This resulted in a modular organised value chain. HK was one of the major destinations. Problems of low planning accuracy and quick response were shifted from Western firms to Asian firms, which could better deal with the requested flexibility.

4.3. THE ELECTRONICS INDUSTRY IN THE GPRD AND MODULAR VALUE CHAINS

After giving a short introduction to the institutional and general economic setting in the GPRD, this section will outline how the electronics industry in the GPRD has developed (see Section 4.3.1) and what global impact it has (see Section 4.3.2).

Reasons for the rising impact of the electronics industry in the GPRD will be discussed in Section 4.3.3 under consideration of the unique economic setting.

4.3.1. Development of the Electronics Industry in HK and the PRD

The HK electronics industry has undergone substantial changes in the past few
decades. Evolving from the manufacturing of simple items like torches, electronic
toys and land line telephone sets, the electronics industry took off in the 1970s
when semiconductors started to be widely applied to electronic circuits. As described above, in the 1980s and especially in the 1990s Western LFs responded to
rising domestic labour costs and increasing competition by outsourcing the production to CMs in HK, Taiwan or Singapore. CMs mainly produced on an OEM basis.
HK benefited from this trend for three reasons. Firstly, the HK government started
to promote domestic entrepreneurship because of its lack of substantial natural resources. Secondly, Western LFs looked for cost-efficient production in Asia, but
also needed institutional safeguards for their business. HK, as a British colony, provided a functional legal system as a safe trading platform. Thirdly, leading Western
firms were unfamiliar with local conditions in Asia, especially in China. They typically preferred to contract with first tier trading organisations in HK that handled
local bureaucratic procedures and further subcontracted orders to smaller second
tier producers in China. With an increasing demand for electronics products on the
part of HK firms, there was a shortage of factory space in HK.

In the first phase of development in the 1980s, China started to attract FDI by
opening the country for FIE and providing them with reduced taxes (see paragraph
liberalisation, privatisation, industrial and tax policy). HK CMs took advantage
of the opportunity to shift production to the Chinese mainland. Taiwanese and
international CMs followed this path. This large scale relocation of production
lengthened the commodity chain and led to a large number of greenfield factories
in the first development phase 1979–1990 (CARNEY 2005:344–345). The primary
destination of investments was SZ, and later DG, because of their close proximity
to HK. The lack of developed infrastructure made the proximity necessary, because
personnel had to travel between HK and the PRD on a regular basis. HK firms
concentrated on the production and assembly of simple electronic devices (THE
GREATER PEARL RIVER DELTA BUSINESS COUNCIL 2007).

In the second development phase 1991–1996, this trend was strengthened by the
immense demand for electronics items and the improvement of China's institutional
environment (for example the allowance of wholly-owned FIEs, legal reforms). Foreign firms, particularly those other than HK and Taiwanese firms, set up factories in
the PRD. Because infrastructure such as motorways and ports was developed, firms
set up factories in neighbouring cities such as GZ, Foshan, Zhongshan and Zhuhai.

The third development phase 1997–2000 was influenced by HK's return to
China in 1997, followed by the Asian Financial Crisis 1997–1998. By that time,

electronics firms had started efforts to add value to their products and to take over more functions in the value chain (ENRIGHT et al. 2005). A survey conducted by the HONG KONG TRADE DEVELOPMENT COUNCIL RESEARCH DEPARTMENT (HKTDC) (2000b) in HK in 1999 showed that although 82% of the companies surveyed were engaged in OEM production (Original Equipment Manufacturing), 62% identified themselves as being involved in ODM production arrangements (Original Design Manufacturer) and 36% also had original brand name production (OBM). The total does not add up to 100% because some companies may have engaged in more than one type of production arrangement. In fact, the trend is towards branching out to ODM and more particularly to OBM.

In the next development phase 2001–2006, firms in the GPRD continued their upgrading strategy. A repetition of the same survey in 2003 revealed that the share of firms engaged in ODM (63%) and OBM (41%) had increased (HONG KONG TRADE DEVELOPMENT COUNCIL RESEARCH DEPARTMENT (HKTDC) 2003b). With its accession to the WTO in 2001, China had to further improve business conditions for FIEs. A major problem was the low protection of IPRs (SCHUELLER 2002; YEUNG 2002). The central government was confident of making the right improvements and established high-tech parks. FIEs and local firms started to set up their research and development (R&D) centres, and although firms still complained about the inadequate protection of IPRs, they continued to carry out innovation activities using their own production systems. HK engineers trained their Chinese colleagues. The demand for engineers—especially electrical engineering technicians—grew rapidly. Firms had to—and still have to—deal with a shortage of engineers. Knowledge transfer and problems of copying driven by a high fluctuation of labour became problematic for electronics firms. Moreover, the WTO rules demanded that China open its service sector. HK firms can now enjoy greater support from HK services (for example legal services) in the PRD.

The fifth phase started in 2007 and is characterised by a strict upgrading strategy on the part of the Chinese government. The incentives of the processing trade policy were rapidly reduced. Electronic firms are affected by this because cheaply imported components of electronics products are no longer subject to VAT rebates. The new labour contract law and the new enterprise tax law equate foreign and domestic firms in the PRD. Additionally, the Chinese government hopes to developed the western provinces in its 11th five year plan. Production of simple electronic items, like torches, electronic toys or land line telephones, and pollutive firms (for example battery producers) are being pushed to relocate to the western provinces or upgrade their production activities. The production of high-tech goods, like mobile phones, cameras, laboratory equipment or semiconductors, is subject to incentives. HK firms know that upgrading is their only opportunity for survival in the GPRD. Therefore, some firms concentrate almost exclusively on product development and design. Production activities are frequently carried out by qualified Chinese suppliers. The recent economic crisis has pushed the process for upgrading and relocating, because only firms with a competitive business concept will survive.

4.3.2. Integrating Hong Kong Firms into Modular Value Chains

The global impact of the electronics industry in the GPRD can be studied by looking at the increasing share of trade with electronics products. Table 4.3 shows the export volume and the CAGR of electronics exports to the world. Asia and Europe registered the highest exports of electronics products. However, Asia (20%) had a higher CAGR during the period 2002–2006 than Europe (15%). Exports from the USA are of only marginal importance. On the national level, China reported the highest exports and the highest growth rate (GR) in exports of electronic items. China is responsible for 13% of the world's exports. It also reports a tremendous CAGR 2002–2006 with 38%. The Guangdong province alone counts for 5% of global electronics exports. The rapid growth of exports can be explained by two factors. Firstly, ARAUJO and ORNELAS (2007) found that an improvement of the institutional environment of a country increases trade volume, as it changes the incentives for customers. Secondly, fragmentation and the global scope of the value chain cause trade numbers to rise. Table 4.3 also indicates similar export numbers for Guangdong and HK. As production capacities in HK are reduced to a minimum, the figure suggests that HK is used as a trading hub for goods manufactured in the neighbour province of Guangdong. It can be assumed that end products find their way from the Guangdong province to the world via HK. This can be proven by a detailed analysis of the characteristics of the electronics industy in the GPRD. It can be expected that large HK-owned CMs provide the gateway to the global market, although they have set up their production activities and their supplier networks in the PRD.

Empirical evidence for the development of CMs as a sign of global integration and modularisation of the electronics value chain can be found in the annual publication of REED BUSINESS INFORMATION (n. d.) about the *Top 100 Contract Manufacturers*. Table 4.4 emphasises the development of CMs from 2003 to 2007, but places special focus on the contribution of HK-based CMs. The table shows the increasing impact of Asian-based CMs for the electronics industry. In 2003, only 13 CMs of the TOP 100 were based in Asia, but by 2007, this number had increased to 33. Now, one third of the TOP 100 CMs of the electronics sector are based in Asia. In 2007, 10 of the world's TOP 100 CMs were HK-owned, with 2 having their origin in China. This shows the impact of HK entrepreneurs on the world's electronics platform. Taking the annual turnover of CMs into account, the CAGR (2003–2007) of Asian CMs was 65%. HK CMs had an even higher CAGR (91%). Although HK firms account for 10% of the world's TOP 100 CMs in the electronics industry, they are only responsible for 6% of the annual turnover. HK CMs can be characterised as small on an international scale. It can be confirmed that HK firms

Table 4.3.: Exports in billion USD and growth rates of electronics products to the
world

	2002	2003	2004	2005	2006	GR 2005–2006	CAGR 2002–2006
Exports of electronics products[a] in billion USD							
World	1,877	2,177	2,655	2,968	3,384	14%	16%
Asia	671	832	1,065	1,221	1,398	15%	20%
Europe	795	921	1,107	1,217	1,385	14%	15%
N-America	318	325	368	400	443	11%	9%
China	123	183	264	347	447	29%	38%
Guangdong[b]	–	–	–	136	171	25%	–
HK	94	112	138	162	183	13%	18%
Japan	197	226	275	278	289	4%	10%
GER	209	242	304	337	380	13%	16%
USA	282	287	325	351	390	11%	8%
Share of world's exports of electronics products							
Asia	36%	38%	40%	41%	41%		
Europe	42%	42%	42%	41%	41%		
N-America	17%	15%	14%	13%	13%		
China	7%	8%	10%	12%	13%		
Guangdong[b]	–	–	–	5%	5%		
HK	5%	5%	5%	5%	5%		
Japan	11%	10%	10%	9%	9%		
GER	11%	11%	11%	11%	11%		
USA	15%	13%	12%	12%	12%		

[a]Electronics products according to HS 2002 (see Appendix A)
[b]no data available for 2002–2004

Source: Own calculation based on UNITED NATIONS COMMODITY TRADE
        STATISTICS DATABASE (n. d.), GUANGDONG PROVINCIAL BUREAU OF
        STATISTICS (2007)

take over the role of CMs in the modular organised value chain of electronics products. The upper end of the value chain (LFs to CMs) is well covered in the academic
literature. What is still underrepresented is how the lower end of the value chain is
organised (CMs to CSs).

When looking at the lower end of the value chain, SMEs in HK mainly work on
an OEM basis. 30% of the firms surveyed indicated relying 100% on OEM production, while another 46% of the firms produce more than 50% of their output on an

Table 4.4.: Impact of HK based CMs for the global electronics industry 2003–2007

	2003	2004	2005	2006	2007	CAGR
Number of CMs (TOP 100) in						
Asia, thereof	13	16	23	30	33	26%
Hong Kong	4	7	9	9	10	26%
China	0	0	0	2	2	–
Annual turnover of CMs (TOP 100) in 100 million USD						
in total	681	707	1,073	1,324	1,998	31%
Asia, thereof	180	201	501	736	1,349	65%
Hong Kong	8	18	31	36	110	91%
China	0	0	0	6	7	–

Source: Own calculation based on REED BUSINESS INFORMATION (n. d.)

OEM basis (see Table 4.5). SMEs in HK seem to serve CMs as producers, getting clear directions as to what and how to produce. The picture changes significantly ($\chi^2 = 41.51, p = 0.000***$) when turning to the PRD firms. 35% of the PRD firms indicated having only ODM or OBM production. Fewer firms than in HK make their profits with OEM. This certainly shows that the modular value chain pattern does not work in every case. Some firms in the PRD seem to break through the chain by producing and designing their own products. Another reason for this result, which fits well with the modular value chain pattern, could be that PRD firms are specialised in simple production. They produce items such as conductors, wires or other simple electronic components. For those components, no blueprints are necessary for the production. They have the freedom to design and produce those components. Therefore, they indicated working on an ODM or OBM basis. There seem to be firms which fit well with the modular value chain pattern, but there are also firms which do not. Although the general integration of GPRD firms into modular value chains could be proven, one has to keep in mind that this might not be the case for all firms. The next sections will examine the organisation of the lower end supply chain in more detail.

4.3.3. Unique Locational Setting of the GPRD Region

Trading Links between Hong Kong and the PRD

The special nature of the GPRD does not only refer to its global impact for the electronics industry (see Section 4.3.2), but also to its unique locational setting. It has been shown that Guangdong is responsible for 5% of the world's exports of elec-

Table 4.5.: Manufacturing position of HK and PRD firms in the value chain

	OEM = 100%	OEM > 50%	ODM/OBM > 50%	ODM/OBM = 100%	in total
HK firms ($n = 102$)[a]	30%	46%	8%	16%	100%
PRD firms ($n = 194$)	28%	22%	16%	35%	100%

Note: For the entire thesis stars mark the significance of a result. Thereby, (*) indicate the significance on the 0.1 level, (**) indicate the significance on the 0.05 level and (***) indicate the significance on the 0.01 level.

[a] $\chi^2 = 41.51, p = 0.000^{***}$

Source: Calculation based on own survey conducted in DFG SPP 1233 [2007]

tronics products, but it also has to be expected that most of them find their way to the global market via HK. This can be explained by the existence of tax-free import and export regulations for several goods between HK and the PRD. The way products manufactured in the PRD reach final markets is illustrated in Table 4.6. The Guangdong province is responsible for 31% of the Chinese exports. 16% of the products are exported to HK. 95% of the HK exports are actually re-exports. It can be assumed that most of the products exported to HK are produced in Guangdong and account for most of the re-exports. A look at two products—cordless telephones and electronic calculators—shows that for electronics products, the links between HK and Guangdong are even closer. 97% of the telephones and 64% of the calculators exported from China actually come from the Guangdong province. 60% of the telephones and 33% of the calculators are exported from China to HK, most of them probably originating in the Guangdong province. For both electronic devices, HK reported re-exporting 100% of the items. The local specialisation—HK as a long-term and safe trading platform combined with Guangdong as a cost- and time-sensitive location for production—is pivotal for the success of the GPRD in the global electronics industry. Although Guangdong has improved infrastructural capacities to ship products to the world (expansion of ports in GZ and SZ), a major proportion is still freighted from HK. A study conducted by the FEDERATION OF HONG KONG INDUSTRIES (FHKI) (2003) confirms these results empirically. 67% of the firms surveyed in the PRD indicated exporting all products via HK. 15% reported exporting selected goods to HK, and only 18% of the firms said that HK is of no importance for their export channels.

   Although the PRD is known as a production and export area for electronics products as shown above, it is not able to produce all the varieties of components needed for end products, due to the limited capabilities of firms. High quality com-

Table 4.6.: Trade flow of selected electronics products in million USD in 2006

	in total		Wireless Telephones[a]		Electronic Calculators[b]	
Chinese exports to the world	969,070	100%	1,777	100%	814	100%
thereof exports from Guangdong	301,948	31%	1,724	97%	525	64%
HK imports from China	153,402	16%	1,060	60%	267	33%
HK exports to the world	316,463	100%	1,407	100%	313	100%
thereof re-exports	299,165	95%	1,407	100%	313	100%

[a]HS 2002 code 851711
[b]HS 2002 code 847010

Source: Own calculation based on UNITED NATIONS COMMODITY TRADE STATISTICS DATABASE (n. d.), GUANGDONG PROVINCIAL BUREAU OF STATISTICS (2007)

ponents, such as semiconductors, integrated circuits and printed circuit boards, must still be imported from countries including Korea, Japan, Taiwan, the US, or from Europe. The communist party is pushing the development of high-tech industries, especially the development of semiconductors, in its 11th five year plan, but the quality has still not reached international standards. Although high-end electronic devices are exported from the PRD, integrated high-tech components originally come from developed economies and are only assembled in the PRD. This pattern can be proven using the revealed comparative advantage (RCA) analysis. The RCA analysis relates the export-import-relation of one subsector and the export-import-relation of all sectors in one region. A positive RCA value implies a comparative advantage in the production of an item, because exports are higher than imports. A negative value demonstrates a relative comparative disadvantage in the production of a certain item. The higher or lower the values of RCA, the higher or lower the comparative advantage is (SCHAETZL 2000). Table 4.7 presents the RCA-values for selected electronic sectors and products in China and Guangdong. Air conditioners, colour TV sets (low-tech) and data processing equipment, for example PCs, monitors, printers (high-tech), have a positive value, which confirms the expertise of firms in China and Guangdong for producing and exporting lower- as well as higher-end products. A comparison of the two regions shows that Guangdong has higher RCA-values for air conditioners and TV sets and a lower value in data processing equipment. This makes it clear that firms in Guangdong still concentrate

Table 4.7.: Revealed comparative advantage (RCA) value for selected product categories in China and Guangdong 2006

| | RCA-value[a] | |
	China	Guangdong
Boilers, Mechanic Equipments and Accessories (84[b])	33.4	47.5
Air Conditioners (84.15)	215.2	419.6
Data Processing Equipments (84.71)	133.9	109.5
Machinery, Electric Equipments, TV Sets, Sound Appl. (85)	−16.4	−16.3
Colour TV Sets (85.28)	403.9	469.5
Integrated Circuits (85.42)	−180.6	−308.8
Optical, Photographic, Film, Medial Instruments (90)	−79.3	−81.4

[a]$RCA_{ij} = 100 \cdot ln \left(\frac{x_{ij}}{m_{ij}} / \frac{\sum_{j=1}^{m} x_{ij}}{\sum_{j=1}^{m} m_{ij}} \right)$

with

RCA_{ij} = "Revealed comparative advantage of subspace i in subsector j

x_{ij} = Export value of subspace i in subsector j

m_{ij} = Import value of subspace i in subsector j

[b]according to HS 2002

Source: Own calculation based on UNITED NATIONS COMMODITY TRADE STATISTICS DATABASE (n. d.), GUANGDONG PROVINCIAL BUREAU OF STATISTICS (2007)

more on the production of lower-end items than Chinese firms in general, assuming that the production of data processing equipment requires more expertise. Both China and Guangdong have a negative RCA value for ICs. This confirms again that China still needs to import high-tech products. Again, the Guangdong value is lower than the Chinese value, which shows that Guangdong lacks high-tech firms in comparison to the Chinese average. STURGEON (2003:12–13) confirmed this finding in his studies. He found that the Guangdong province is predominantly responsible for simple standardised products such as personal computers, mobile phones and consumer electronics manufactured on the Chinese mainland.

Division of Labour between Hong Kong and the PRD

The GPRD region is not only unique in its specific trading pattern within the region resulting from the two different institutional environments, but also because of the division of labour between the two regions. Both regions provide comparative advantages for certain business activities. This leads to an increase in the overall regional competitiveness. In the beginning, the PRD region only held production advantages over HK in the areas of factory space, special tax conditions and cheap labour. But nowadays the division of labour is more specific. The PRD provides excellent conditions for establishing service activities which were formerly located in HK. When looking at customer producer relations, it is of interest where firms in HK and the PRD have set up their sales and procurement activities. Table 4.8 shows where electronics firms surveyed in HK and the PRD *mainly* apply their management, production, procurement and sales activities. The share of firms active in HK and the PRD together does not add up to 100%, as some firms use locations outside the GPRD. The table comparatively presents the responses of firms in the HK survey, the PRD survey (differentiated according to location and ownership of firms) and for validity provement responses of firms from a survey conducted by the FEDERATION OF HONG KONG INDUSTRIES (FHKI) (2003). Firms in HK tend to keep their management activities mainly in HK (72%), because HK usually serves as the regional headquarters. Production activities are located in the PRD (90%). Nearly half of HK firms have their procurement activities located mainly in HK, while the other half has them located in the PRD. This fits the findings that procurement of raw materials and simple components takes place in China, whereas high-tech components are sourced abroad. As products are produced predominately for the international market, HK is known as the "front shop" of the GPRD-region. Therefore, 87% of the HK firms set up their sales activities in HK.

A study published by the FEDERATION OF HONG KONG INDUSTRIES (FHKI) (2003) aimed to identify the close relationship between HK and the PRD by inter-viewing firms in the PRD with links to HK. Although they did not focus on the electronics industry itself, their study produced similar results. The sales and pro-duction figures are nearly identical. The FHKI survey indicates a greater share of firms having their management activities in HK. This can either be explained by the specific focus on the electronics industry, which is more often managed in the PRD than other industries, or by the time difference between the two surveys (2003 and 2007) considering the continuous trend of outsourcing management and service activities (BALDWIN 2006). Firms in the PRD usually carry out their activities in the PRD. But differences between the locations GZ and DG are evident. GZ fig-ures behave similarly to the average, but DG firms more often carry out sales and procurement activities in HK. This can be explained by a higher concentration of foreign firms in DG, whose links to HK are usually closer. Foreign firms are able to localise their business activities where they are of most benefit. For Chinese firms, it involves more capital, expertise and administrative effort to set up business activ-ities in HK. This can be confirmed by comparing the behaviour of HK and Chinese (CN)-owned firms. Whereas Chinese-owned firms keep management and produc-tion activities strictly in the PRD and rarely set up procurement or sales activities in

Table 4.8.: Division of business activities between HK and the PRD

	Unit Location	Management	Production	Procurement	Sales
HK survey	HK	72%	8%	52%	87%
	PRD	28%	90%	48%	13%
FHKI survey	HK	90%	11%	64%	86%
	PRD	10%	89%	36%	14%
PRD survey	HK	4%	0%	5%	15%
	PRD	92%	97%	87%	71%
/Guangzhou	HK	3%	1%	10%	18%
	PRD	93%	93%	83%	70%
/Dongguan	HK	12%	4%	14%	26%
	PRD	82%	95%	74%	58%
/CN owned	HK	1%	1%	7%	7%
	PRD	94%	94%	83%	81%
/HK owned	HK	18%	6%	28%	44%
	PRD	82%	94%	73%	54%

Source: Data based on own survey conducted in DFG SPP 1233 [2007], FEDERA-
TION OF HONG KONG INDUSTRIES (FHKI) (2003)

HK, HK-owned firms more often take advantage of the two different regional settings in the GPRD. This suggests that now, not all firms are able to take advantage
of the division of business activities between HK and the PRD. HK-owned firms in
particular profit from the opportunity to shift some of their activities into the PRD.
Chinese-owned firms' opportunity to benefit from the unique locational setting is
limited due to difficulties in accessing HK.

Predictability of Industry Conditions of Hong Kong and PRD Firms

The different spatial distribution of the business activities of firms surveyed in HK
and the PRD interferes with the ability to gain information for predicting industry conditions related to sales and procurement activities. HK-based firms usually
have their sales division in HK, while PRD firms keep their sales offices in the
PRD. The surveys conducted clearly show that HK firms profit from their infor-
mational edge over firms in the PRD. They could predict industry conditions re-

lated to customer activities (price, volume, quality, delivery time) significantly better ($\chi^2 = 5.83, p = 0.054^*$) than PRD firms in the last five year period (2002–2007). 36% of the HK-based firms indicated that they find changes in price, volume, quality and delivery time very predictable, in comparison to only 30% of the firms in the PRD (see Table 4.9). There are three explanations for this. Firstly, HK firms have a locational advantage in being able to set up their sales office in HK, which provides better access to customers (for example fairs, business organisations). Secondly, HK firms have more experience on international markets and better forecasting systems. Thirdly, HK firms enjoy a better relationship to customers (longer relationships, better reputation, greater reliability). Taking the predictability of producer-related information into account (availability of resources and suppliers, labour, governmental regulations in China), HK firms and PRD firms indicated no significant difference ($\chi^2 = 3.93, p = 0.140$). The data reveal that PRD firms cannot predict sourcing-related industry conditions better than HK firms, although they mainly carry out their activities in the PRD. HK firms' long-term activities in the PRD seem to enable them to forecast industry conditions as well as PRD firms do. For both HK and PRD firms, it is true that the ability to predict changes related to customers is higher than that of changes related to producers (HK: $\chi^2 = 98.24, p = 0.001^{***}$, PRD: $\chi^2 = 43.23, p = 0.001^{***}$). Overall, Table 4.9 indicates that only a few firms are able to predict industry and market changes. This implies a high market volatility and the need for flexibility if firms are to compete on international markets.

To summarise, the electronics firms in the GPRD were responsible for 5% of all global exports of electronic products in 2006. Electronics firms in the GPRD are highly integrated into the modular global value chains. HK hosts 10% of the world's largest CMs. These produce in the PRD and set up their network of component suppliers there. The unique setting of the PRD can be attributed to a clear division of business activities between HK and the PRD. Whereas firms prefer to organise their management and sales in HK, they go for production and procurement activities in the PRD. The PRD is specialised in the export-oriented production of consumer electronic items like telephones or calculators. However, high-tech products such as laboratory equipment and cameras are becoming more important. The sourcing of simple components is usually done in the PRD or China, but high-tech components such as semiconductors are still imported from developed countries. Investigations revealed that HK firms are able to profit more from the unique locational setting in the GPRD. Their access to information, provided by management activities in HK, allows them them to predict market and industry conditions more accurately regarding customer and producer relations.

The market success of firms is not limited to the performance of a single firm, but to an entire production network. Only if the entire network of customers and producers is arranged flexibly are firms able to succeed on the world market. Therefore, the next section explicitly focuses on the spatial distribution of customers and

Table 4.9.: Firm's ability to predict changes of industry conditions related to customers and producers 2002–2007

	Related to customers[ac]		Related to producers[bd]	
	HK sample[e]	PRD sample[f]	HK sample	PRD sample
	$n = 104$	$n = 221$	$n = 104$	$n = 219$
very predictable	36%	30%	20%	16%
moderate predictable	53%	49%	37%	46%
not predictable	12%	21%	43%	38%
in total	100%	100%	100%	100%

[a]price, volume, quality, delivery time
[b]availability of resources and suppliers, labour, governmental regulations in China
[c]HK:PRD variable customer $\chi^2 = 5.83, p = 0.054^*$
[d]HK:PRD variable producer $\chi^2 = 3.93, p = 0.140$
[e]customer:producer variable HK $\chi^2 = 98.24, p = 0.001^{***}$
[f]customer:producer variable PRD $\chi^2 = 43.23, p = 0.001^{***}$

Source: Calculation based on own survey conducted in DFG SPP 1233 [2007]

producers of HK and PRD firms. The location of customers and producers heavily influences the organisational modes applied in customer and producer relations,
which are ultimately responsible for a flexible network of firms.

4.4. SPATIAL DISTRIBUTION OF CUSTOMERS AND PRODUCERS

This section focuses on the spatial distribution of customers and producers of HK
and PRD-based electronics firms. This provides information as to how electronics
firms in the GPRD combine global links and local concentration in production.

To illustrate network activities in the GPRD, firms in HK and the PRD were
asked to provide information about the location of their most important producer
and their major customer as suggested in Section 3.2.1 about operationalisation. HK
firms' most important customers are mainly located in final markets, meaning North
America and Europe (52%). The proportion of Japanese customers is only 8% (see
Figure 4.2). 15% of the HK firms' most important customers come from China.
14% of the products are indicated as being consumed in HK. This high number can
be explained by firms that do not have complete knowledge of their final markets
(for example producers of parts and components). They indicated the final market
as being in HK. In contrast, 41% of firms surveyed in the PRD indicated having
their most important customer in HK. However, their final markets (China 38%,
North America and Europe 23%, Japan 8%, HK 13%) differ substantially from the

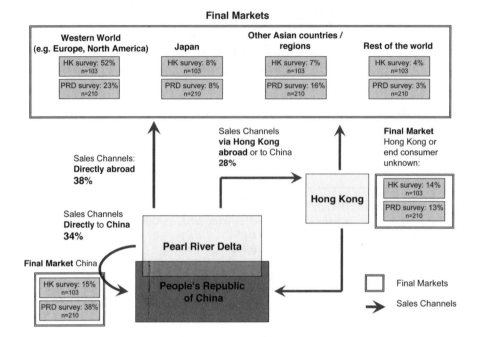

Figure 4.2.: Location of final markets and sales channels of firms in the GPRD
Source: Own draft and calculations based on DFG SPP 1233 [2007]

location of their most important customer. This implies that firms in the PRD only partly work with customers located in the markets of their final products. Before their products reach the final markets, other firms step in. They are apparently at the lower end of the value chain system in comparison to firms in HK. Firms in the PRD are less likely to produce final products. This illustrates that firms in HK play an essential role as a customer for products manufactured in the PRD.

As the location of final markets and the location of the major customers for firms in HK are identical, a direct product flow to final markets is assumed. However, this pattern could not be found for firms in the PRD. To understand how products of firms in the PRD are transferred to their final markets, firms in the PRD had to indicate which sales channels they use. According to the survey in the PRD, 28% of all PRD sales are organised via HK. They are distributed as follows: 13% of all PRD sales are managed by HK-based affiliates, 8% via HK-based trading companies and 7% via other HK-based firms. Products sold via HK are intended either for the Chinese or for the international market. 38% of the PRD sales go directly to international markets and another 34% are sold directly to China. A differentiation according to the ownership structure reveals that HK-owned firms in the PRD are responsible for the biggest share of products sold via HK, whereas Chinese and Taiwanese-owned firms in the PRD account for the high share of products sold directly to the Chinese market. For HK entrepreneurs in the PRD, it is profitable to use their sales and distribution networks in HK, which were established in the 1980s and 1990s.

Firms in HK and the PRD are integrated into the global value chain system. Products are mainly produced for the international markets. LALL et al. (2004) studied the fragmentation of electronics products. They also identified a high export rate of electronics products from HK and China, which shows the global integration of the GPRD. But a distinction must be made between firms in HK and those in the PRD. HK firms tend to occupy a position at the upper end of the value chain, but this is only partially true of firms in the PRD. About one third of sales are made via HK. Another third of sales are not designated for the international market, but rather for China.

In contrast to the international distribution of customers, firms in HK responded that 39% of their producers were located in SZ and the same proportion in DG. Another 13% of producers could be found in first ring cities (GZ, Foshan, Zhongshan, Zhuhai) and 8% within the second ring (Huizhou, Zhaoqing, Jiangmen). The clustering of HK producers in the core cities SZ and DG is caused by large HK firms establishing their first PRD link in those cities immediately after the first Chinese opening. Interviews conducted showed that SZ's and DG's proximity to HK was pivotal, because personnel (CEO, managers, engineers) travel between firms on a regular basis for a period of two to three days a week. Other criteria named were the infrastructural development, the proximity to other firms and personal relationships within those cities. A distribution test for a bias in firms' age, innovativeness and the share of OBM/ODM production in the cities of SZ and DG showed these factors to be insignificant.

Although firms in HK indicated having their most important producers in SZ and DG, the PRD survey focussed mainly on DG and GZ (for reasons see Section 3.1). 41% of the firms in the PRD survey are located in DG and another 53% are located in GZ, while only 6% are located elsewhere in the PRD. In both cities, districts with a high density of electronics firms were selected. The data set shows that in GZ, 61% of the firms are mainly Chinese-owned, while in DG only 41% are. Foreign firms in GZ are most likely to be from HK, whereas a foreign-owned firm in DG is most likely to be Taiwanese-owned. Differences in the ownership structure are significant ($\chi^2 = 21.2, p = 0.001^{***}$). Taking the age of firms into account, firms have been classified according to their foundation year: before 1990, 1991–1996, 1997–2000, 2001–2003 and 2004–2007. Firms in DG were found to be slightly older than in GZ. The share of firms set up in 1990 or earlier differs the most (DG: 12%, GZ: 7%). Comparing firms set up from 2004 to 2007, a reverse figure can be recognised (DG: 14%, GZ: 21%). The bias in age

also turns out to be significant, but only on the 0.10 level ($\chi^2 = 9.1, p = 0.057^*$).
A look at the manufacturing type (OEM=100%, OEM>50%, ODM/OBM>50%,
ODM/OBM=100%) shows that firms in DG are more likely to be engaged in OEM
production. 27% of the firms reported only focussing on OEM production, while
another 31% generate at least half of their sales in OEM production (GZ: OEM
= 100%: 29%, OEM>50%: 14%). The share of firms with ODM/OBM>50% is
16% for both cities. Huge differences are evident when looking at the share of firms
with ODM/OBM=100%. In GZ, 41% of the firms reported being in this category,
whereas in DG the figure is only 27% ($\chi^2 = 9.1, p = 0.028^{**}$). It reveals that in DG,
foreign electronics firms were set up early and concentrated on the OEM production
for exports. In GZ, firms were set up later, but did not focus so much on OEM, concentrating instead on the ODM/OBM business immediately. Although differences
in the type of manufacturing were evident, innovation activities were not found to
be higher in GZ than in DG.

It has been shown that orders placed to HK firms are forwarded to the first tier
producer in the PRD. In order to follow the value chain, PRD firms were asked
to indicate the spatial distribution of their producer network. This second tier or
component producer network is also located in the PRD in proximity to the first
tier producers (see Table 4.10). 16% of the firms in DG reported having their most
important producers in the same city. 19% indicated having their most important
producer in SZ and another 33% reported the producer being located elsewhere in
Guangdong. 13% of the firms work with a producer in China, but outside Guangdong. Producers in GZ indicated having 36% of their producers in in same city
and 24% elsewhere in Guangdong. Firms in DG and SZ are underrepresented. In
DG and GZ, about 17% and 18% respectively of the main producers are located
abroad. It has already been shown that high quality products such as semiconductors are sourced abroad (see Table 4.7). This is confirmed in the data set as well.
For component sourcing other than for technologically very sophisticated products,
the producer networks in the PRD seem to be locally concentrated within the PRD
without far-reaching interweavement. A comparison between firms in DG and GZ
shows that producers of firms in GZ are significantly more spatially concentrated
than producers of firms in DG ($\chi^2 = 66.5, p = 0.001^{***}$). Firms in GZ are mostly
small and Chinese-owned. They are apparently prone to local clustering. Another
explanation is that DG was developed first. Firms there can be expected to have
more interrelations with firms in other areas of the PRD. Additionally, firms in DG
are more often foreign-owned, which provides them with an extended network.

DG and GZ were selected as research areas because differences in the connections to HK are expected. Table 4.11 compares the different types of HK connection
of firms surveyed in DG and GZ. In total, 65% of the firms surveyed in the PRD reported having at least one HK connection. A distinction between GZ and DG shows
that more firms in DG (78%) are connected to HK than firms in GZ (53%). Having

Table 4.10.: Geographical proximity of firms in the PRD and their most important
            producers

Location of producers[a]	if firm in Dongguan $n = 108$	if firm in Guangzhou $n = 83$
Same city	16%	36%
Guangzhou/Dongguan respectively	2%	5%
Shenzhen	19%	7%
Guangdong (elsewhere)	33%	24%
China (elsewhere)	13%	10%
abroad	17%	18%
total	100%	100%

[a]$\chi^2 = 66.5, p = 0.001^{***}$

Source: Data based on own survey conducted in DFG SPP 1233 [2007]

the most important customer in HK was the most frequently indicated HK connection (40%). HK's role as a major supplier for components proved to be unimportant (5%), but 27% of all firms source or import at least some of their components from HK, especially high-tech goods. Of secondary importance are the sales relations to HK (32%). DG firms' share thereof is higher (39%) than the GZ firms' share (25%). HK is of major importance for innovative products in particular. Additionally, 18% of the PRD firms are HK-owned, with the share in DG being slightly lower than in GZ due to the fact that firms in DG often come from Taiwan (YANG 2007). Although the share of HK firms is higher in GZ, DG hosts more foreign firms overall (63% in GZ and 45% in DG). This also explains why a larger share of firms in DG indicated using HK-based affiliates or traders as sales channels to final markets. To conclude, firms in DG have stronger and more diverse connections to HK than firms in GZ. For both cities though, HK firms act mainly as customers or sales agents.

To conclude, the surveys showed that HK firms mainly use a global customer base. In contrast, PRD firms focus on both the Chinese and the international market. PRD firms sometimes deal directly with their customers in international markets but also use HK firms as transmitters because of their own limited sales network. FEENSTRA et al. (2002) compared Chinese firms that export through HK to those that do not do. They calculated that Chinese firms profit from exporting through HK due to a concentration of the comparative advantages. A detailed view of the production networks of HK-based firms in the PRD shows that HK firms' producers are mainly located in DG and SZ. HK firms were the first to move their production

Table 4.11.: Type of HK connection of firms in the PRD

| | Percentage of firms | | |
	in total $n = 222$	Guangzhou $n = 116$	Dongguan $n = 89$
Any HK connection	65%	53%	78%
No HK connection	35%	47%	22%
Main customer in HK	40%	29%	53%
Main supplier in HK	5%	8%	2%
HK as market for sales	32%	25%	39%
HK as market for sales of innovative products	22%	14%	33%
Sourcing/Import from HK	27%	25%	30%
Firm is mainly HK owned	18%	21%	15%
Sales/Marketing unit is located in HK	19%	15%	26%
Logistics/Distribution unit is located in HK	15%	13%	19%
Financing unit is located in HK	11%	7%	17%
Procurement unit is located in HK	10%	9%	12%
Management unit is located in HK	6%	3%	10%
Innovation activities are located in HK	4%	2%	7%
Production unit is located in HK	2%	1%	3%
HK as sales channel: via HK bases affiliates	18%	14%	25%
HK as sales channel: via HK based traders	14%	8%	19%
HK as sales channel: via other HK based firms	9%	8%	9%

Source: Data based on own survey conducted in DFG SPP 1233 [2007]

into the PRD and concentrated on these two cities. Although HK firms have a relatively low preference for GZ, a comparative survey of firms in DG and GZ shows that producers in GZ also have strong indirect connections to HK. The network of second tier producers in DG and GZ is mostly located in close proximity. Despite the extensive international distribution network of customers, one characteristic of the production network is its local concentration.

4.5. SUMMARY AND DISCUSSION OF FINDINGS IN SECTION 4

It could be shown that the global electronics industry has undergone major changes since the 1990s. A growth in demand, shorter product life cycles, modularisation and outsourcing activities have changed the chain structure from a hierarchical to a modular value chain. Western LFs outsourced production activities to CMs in developing countries in Asia. HK firms were favoured because HK offered an appropriate institutional setting to safeguard contracts while organising low-cost production activities in the PRD. HK's CMs increased their impact rapidly. In 2003, HK hosted four of the 100 TOP CMs in the electronics industry, and by 2007 the number had increased to ten. The network of CSs of HK firms is mainly located in the PRD. Proximity facilitates exchange. Hypothesis A3—*HK and PRD firms integrate themselves into modular value chains*—can be confirmed. Large HK firms function as CMs, whereas small or medium sized firms in the PRD take the role of CSs. HK firms build their producer networks mainly in the PRD, with SZ and DG being of greatest importance. Following the value chain, firms in the PRD also prefer having their producers in geographical proximity. About 70% of the PRD firms' producers are located in the PRD. Only some high-tech components are sourced from producers abroad.

In contrast, HK firms have their most important customers located in their final markets. PRD firms partly serve HK firms and partly reach the same position in the value chain as HK firms, dealing on the global market. This can also be confirmed by the appearance of the first Chinese-owned CMs. Hypothesis A5—*the production system relies on a global customer base and a spatially concentrated producer network*—can only be partly confirmed. It is mainly HK firms which have globally spread customers and locally concentrated producers. For PRD firms, customers are partly located in HK and partly firms also deal directly with overseas customers. Additionally, components are not exclusively sourced from China, but also from abroad.

The economic growth of the Guangdong province due to outsourcing tendencies of Western LFs could be shown by an immense annual growth during the period 1980–2005 of FDIs (20%), GIO (15%) and trade volume (17%). In 2006, electronics firms in Guangdong exported 5% of the world's electronics products. The GDP per capita in Guangdong rose to 2,882 USD in 2005. HK firms have had an enourmeous impact for the economic growth of the PRD. They were responsible for 47% of the FDI and 30% of the GIO in 2006. Hypothesis A1—*electronics firms in the GPRD increased their market share in a growing industry sector due to outsourcing tendencies of leading Western firms*—cannot be rejected. But on a city level, it could be shown that some cities in the PRD profited more than others.

Hypothesis A4—*the GPRD region profits from a clear division of labour*—could be proven for HK firms. They organise their management and sales activities in HK, but have moved production and procurement activities to the PRD. In contrast, the business activities of Chinese firms are more locally concentrated. They hardly profit from the division of labour. The majority of HK-owned firms capitalise on the unique locational setting and feel the freedom to set up units where they are most likely to be profitable.

Hypothesis B1—*China's formal institutional environment still has weaknesses concerning the safeguards of trade between firms*—could also be proven. Although major reforms have been applied and improvements achieved, firms still complain of insufficient protection of property rights, especially IPRs, and the restricted opening for legal and financial services.

The global integration of the electronics industry in the GPRD seems to have allowed firms to meet the requirement of flexibility. Product life cycles have shortened, leading to the need for more frequent changes to product features and cosmetic design in order to lure consumers. This makes predictability and planning difficult. It could be shown that HK firms can predict customer-related information significantly better than PRD firms, but the majority of firms still indicated only moderate or zero predictability. Producer-related information was even more difficult to predict for both HK and PRD firms. Hypothesis A2—*low predictability of market-related information*—cannot be rejected. Market volatility is strong and predictability low. Although HK-based firms still have access to more and better customer-related information, planning accuracy is still low and quick response is required by most of the firms. Moreover, CMs have to fulfil the high demands for product variety and quickly changing collections and series. They are in competition to receive orders from international customers. CMs can then pick the appropriate CSs in the PRD, which may be more specialised. To conclude, firms in the GPRD are asked to provide LFs with a certain degree of quantitative and qualitative flexibility (see Figure 2.4). This confirms Hypothesis A6.

   G *How are electronics firms in HK and the PRD integrated into global value chains? How is the division of work spatially organised between HK and the PRD?*

   H *How well is the formal institutional environment in HK and the PRD developed? What informal institutions guide the way of doing business in HK and the PRD?*

Finally, research questions **G** and **H** can be answered. It could be shown that firms in the GPRD are integrated into global value chains. Their global impact is increasing. With integration into the global system, firms must respond to the flexibility required by global customers. Firms in the GPRD succeed in doing so. The two different institutional systems facilitate a clear division of labour. HK functions as a front shop with institutions built on an international level, whereas the PRD is in the position of a back factory providing good production conditions, but still lacks institutional safeguards. However, geographical and cultural proximity help firms respond to the need for flexibility. Moreover, this pattern makes outsourcing to the GPRD safe and cost-efficient for Western LFs. But it has to be kept in mind that the division of work is changing. The central government encourages firms to upgrade their business activities or relocate. Moreover, China has opened for the service sector. Whereas the PRD will specialise on high-tech production, HK's challenge is to concentrate on specialised and extravagant services for those enterprises. The region will remain important for the international economy in future, but on a different level.

It could be shown that HK and Chinese electronics firms are integrated into a modular value chain. Large CMs are located in HK and a few of them in China. The theory of global value chain analyses the relationship between Western LFs and CMs in detail, but gives few clues as to how the CS network is organised. This chapter shows that CSs usually cluster in close proximity to CMs. This seems to be especially crucial when the institutional setting does not provide enough safeguards and firms must compensate for the lack formal rules by way of informal safeguards requiring geographical and cultural proximity. The next chapter will analyse more accurately which governance modes are applied between HK firms and their producers in the PRD to provide more information about how the production network is organised, taking the two institutional settings into account. This should shed light on the lacking discussion about the organisation of CSs at the lower end of the value chain. The value chain could be modular at the first level (LFs—CMs in Asia) but differently organised at the second level (CMs—first tier producers) or at the third level (first tier producer—second tier producer). The electronics value chain could be characterised by different governance modes regulating the relationships between customers and producers on different levels. The conceptual discussion in Chapter 2.4 showed that governance modes are associated with different types of institutional settings. Which governance modes characterise the lower end of the value chain is examined in the next section. It is expected that governance modes differ according to the location of customers and producers. This implies that the value chain might not be purely hierarchical or market-oriented, but can switch modes according to its environment.

5. GOVERNANCE MODES AND CHARACTERISTICS IN CUSTOMER PRODUCER RELATIONS

This chapter focuses on the applied governance modes and their relations to the institutional environment and special firm characteristics. It addresses research question **I** and investigates:

- whether HK firms organise their business relationships to customers and producers in China differently from those to their customers and producers abroad

- whether this a consequence of the different development stages of the institutional environment.

First of all, the applied governance modes in customer producer relations of HK firms as well as those of PRD firms will be focussed on in Section 5.1. In order to prove the formulated hypotheses, different variables will be tested as to their significance for the choice of governance modes. Two logit models have been applied in Section 5.2 to prove the association.

The first model was selected to show whether the institutional environment influences the organisation of relationships to customers and producers. The distinct institutional settings in China and abroad may have a statistically significant influence on the choice of governance modes, as suggested by economic theories. This model concentrates more on relationship-specific variables of customers and producers, rather than on the structure of firms. The second model focuses on the PRD producers of HK firms in order to show (1) the flexibility applied in the different governance modes and (2) the change of governance modes in recent years. Here, the structure of firms is taken into account. It is predominantly the governance modes of HK firms which are studied, because they seem to be of more value in testing the theoretically supported hypotheses. Finally, results are summarised and an answer to the research question is given.

5.1. APPLIED GOVERNANCE MODES IN CUSTOMER PRODUCER RELATIONS IN HONG KONG AND THE PRD

This section briefly concentrates on the applied governance modes towards customers and producers, firstly for HK firms and then for PRD firms. According to the theoretical discussion in Section 2.4.4, firms in HK are expected to organise their relationships to producers or customers in the PRD in a hierarchical-related construction because of the incomplete institutional setting. In contrast, relationships to international producers or customers are assumed to be organised on a market basis. Section 4.4 outlined that firms in HK tend to be linked to producers concentrated in the PRD, but have customers spread internationally. Firms in HK were explicitly asked to indicate which governance mode they use to organise their relationships to their most important customer and producer respectively.

Indeed, for 55% of the HK electronics firms the most important producer is a wholly-owned affiliate in the PRD (see Table 5.1). In the 1980s and 1990s, when most HK firms shifted their production to the PRD, firms feared high legal uncertainty. In the case of conflicts with market-related producers, the production process could be negatively affected. Even though the situation has gradually improved over time, for example following China's accession to the WTO in 2001, firms have generally stayed with their own production plants. Another 16% of HK firms hold shares in their PRD producers in the form of equity cooperation (EC). That way, HK firms still have the power to influence and direct their PRD producers. From 1978–1986, it was not allowed to register a completely foreign-owned firm. Foreign firms were forced by the central government to take on a Chinese partner. Even though this is related to an earlier period, the history of dependency is still recognisable in remaining JVs of HK and Chinese entrepreneurs. Besides, JVs still provide competitive advantages for certain business activities. In addition, some HK firms are targeting a long-term relationship with their producers indicated by framework agreements. It is a way to connect more closely to producers and to share business targets as well as information without equity relationships. Nonequity cooperation (NEC) between HK firms and PRD producers could be observed in 16% of cases. In comparison, only 13% of HK electronics firms work through a pure market relationship (buying and selling agreements) with their most important producer. This study reveals a mostly hierarchically organised cross border production organisation. WEI et al. (2004) studied entry modes of foreign firms in China. They also found that hierarchies are the first choice when foreign firms enter China for the first time. But KRUG and HENDRISCHKE (2006a) claim that cooperative forms have also become more common in countries with incomplete but improving institutions.

A look at the governance modes of customer relations of HK firms provides a completely different picture. 66% of the HK firms indicated having only buying and selling agreements with their most important customer (see Table 5.1). Another 28% work on a NEC basis with their most important customer, and only 6% have any equity relationships at all. HK foreign trade is protected by a complete set of institutions dealing with the conclusion and enforceability of trading contracts and the exchange of financial capital. This decreases transaction costs on markets.

Firms in the PRD produce for international consumers, but as Figure 4.2 shows, they use HK as a gateway to the world. Indeed, 40% of the firms surveyed in the

Table 5.1.: Governance modes applied in customer and producer relations of firms in HK and the PRD

Firms in	most important	Hierarchy	EC	NEC	Market	total
HK	Customer	2 (02%)	4 (04%)	29 (28%)	69 (66%)	104 (100%)
	Producer	57 (55%)	17 (16%)	17 (16%)	13 (13%)	104 (100%)
PRD	HK Customer	16 (18%)	9 (10%)	6 (07%)	57 (65%)	88 (100%)
	Producer	20 (09%)	13 (06%)	42 (20%)	140 (65%)	215 (100%)

EC=Equity cooperation, NEC=Non-equity cooperation

Source: Calculation based on own survey conducted in DFG SPP 1233 [2007]

PRD indicated having their most important customer in HK. As this study focuses on the links between HK and the PRD, the following analysis concentrates only on the links between PRD firms and HK customers, rather than taking all customer relationships into account. Electronics firms in the PRD survey mainly indicated working on a market basis with their customers in HK (65%). Only 7% of the firms work in NEC and 10% in EC. 18% of the firms surveyed are affiliates of HK-based firms. Although over half of all firms surveyed in HK have their own affiliates in the PRD, they only account for a small proportion of all firms in the PRD. The PRD firms which indicated being Chinese-owned tend not to have an affiliate in HK, whereas HK-owned firms generally do. When firms in the PRD survey indicated how they organise their relationship to their most important producer, 65% reported buying supplies from independent producers. Another 20% work under NEC and only a few of them have any equity relations to their producers. Producers are mainly located in the same district.

WANG and NICHOLAS (2007:131–137) also confirmed in their article that NEC has become more important for organising relationships in the Guangdong province. Moreover, HOWARD and SQUIRE (2007) provide support for the notion that product modularisation will lead to greater levels of customer-producer collaboration.

5.2. CHARACTERISING GOVERNANCE MODES

Section 5.1 showed that governance modes of HK firms differ between customer and producer relations. Section 4.4 showed that the locations of HK firms' customers and producers differ. This section concentrates on the analysis of characteristics of governance modes using two logit models. According to the theoretical discussion, it will be tested whether or not the governance modes of customer and producer relations is determined by the institutional setting, as stated by WILLIAMSON (1998). Besides that, other determinants are tested in order to analyse how strong the effect of institutions is. Moreover, a second logit model is used to establish whether the improvement of the institutions and other structural variables affect the applied governance modes.

5.2.1. Governance Modes and Institutional Differences: Logit Model

In Section 2.4 the traits of governance modes and the influence of the institutional environment have been discussed. Firstly, Section 2.4.4 outlined that firms tend to choose a hierarchical form of governance when the *formal institutional setting* does not provide sufficient safeguards to protect the business. Firms choose market relationships when business safeguards are institutionalised in laws and regulations. WANG and NICHOLAS (2007) recommended not only an emphasis on the two extremes, but also a focus on cooperation. As cooperation is manifold, they advised distinguishing between EC and NEC. It is expected that HK firms are more likely to organise their relationships to international partners in a market-like way than firms in China, which prefer hierarchical relations. Secondly, Section 2.4.3 discussed the *transaction-specific environment* as a determinant for governance modes. Frequent transactions involve the lowest costs in hierarchies, whereas rarely-made transactions are cost-efficient in market relationships or NEC. Additionally, the uncertainty and the relationship-specific investments are associated with certain governance modes.

To clarify how important the institutional environment is in determining the governance modes, links with both international and Chinese customers and producers are taken into account. The focus is set here on relationship-specific variables. The relationship to the most important customer and producer was analysed separately in this model. This doubles the number of cases considered in the first logit model. It is similar to a survey in which the number of firms is doubled, with half of the firms providing information about their producer and other about their customer relationships. To prove the significant influence of the institutional setting for the organisation of governance modes of HK firms, three important issues have to be discussed.

1. It can be assumed that decisions on governance modes are highly consequential for HK firms, as they cannot be revised or changed easily. Therefore, it is expected that predominately formal rules are taken into account for decision-making. As the formal institutional framework depends on the location, it is expected that the location of partners has a large influence on the choice of governance modes. Following WEI et al. (2004), this study also takes the location of firms as an indicator for a special institutional environment. The location of customers and producers is expected to be a determining factor for the selected governance mode. The impact of locations must be compared to other factors in order to exclude their influence. This finally confirms the

assumption that institutional factors are most responsible for the selected governance mode.

2. There is no evidence that specifications in the electronics industry generate differences in the organisation of customer and producer relations. Therefore, it can be expected that a firm being a producer or a customer has no influence the governance mode. Even though the majority of customers for HK firms are located abroad and the majority of producers are located in China, the differences in governance modes depend predominantly on location and not on whether a firm is a customer or a producer.

3. This legitimises the method of combining customer and producer relations in a logit analysis to analyse the influences of locations on the governance modes. For the following logit model, governance modes towards partners are studied without distinguishing between customers and producers. It is normally, of course, a doubtful method to combine customer and producer relations, because influences on the governance modes specific to customers and producers (for example position in value chain, power asymmetries etc.) are thereby ignored. The second model will focus on the firm-specific influences on governance modes. But despite some doubts as to the method, combining the two provides a good opportunity to show empirically whether firms really tend to adapt their governance modes to the institutional environment. Because it is difficult to collect data from firms operating in different institutional environments, there is virtually no empirical evidence which could legitimise an attempt to test differences in governance modes in this way while taking care to interpret results.

To clarify the characteristics of governance modes, the logit model tests the influence of *location of partners* as a dummy variable (1=China, 0=abroad). But it also includes other potential factors to be linked with certain governance modes (see Table 5.2). Besides the location of partners, the location of HK firms' unit of collecting information and selecting partners is included to see whether the location of decision-making also has an influence (dummy variable: *Unit is operated in HK*). Additionally, a test was conducted as to whether the *unit is linked to outside supporters* (for example trading firms, consultancies, universities) to gain better information and optimise their decisions on a governance mode. The important role played by guanxi in China has also been discussed. To show whether this fact influences the choice of governance mode, the logit model includes a variable indicating whether the unit dealing with customers or producers respectively is *led by a family member*. Another relevant variable included is the need for flexibility resulting from the market pressure in the electronics industry. In order to deal with low *predictability of supplies and sales* (Likert scale: 1=very predictable to 5=not predictable), firms may organise their relations differently. Apart from the firm-specific characteristics, the model also tests the influence of relationship-specific characteristics linked to certain governance modes. It has been mentioned that governance modes also depend on the transaction-specific environment (frequency, uncertainty and specific investments). Transactions which are repeated frequently are more often organised in hierarchies. The length of *working experience in years* is taken as an indicator to show repeated transactions. Moreover, the share of products sold or sourced from a partner in value terms can be taken as a sign for relationship-specific investments. If the *dependency* of partners is high, it can be expected that investments in this relationship have been higher. This is assumed to go along with hierarchies or EC, which naturally require higher investments. To measure the quick response ability of firms in certain governance modes, the *time of negotiation in hours* is included. This is the length of time from when orders are placed until all details are figured out and the production process can start.

Not all variables can be taken as determinants of governance modes. Determinants are obvious before a governance modes is selected (for example location of partners, location of decision-making unit, unit led by family members, etc.). But some other variables only become obvious once the mode has already been selected (for example working experience, dependency, time of negotiation). This approach follows the transaction cost theory. It outlines that governance modes are related to transaction-specific characteristics such as frequency or specific investments. Those variables are also unknown in their specification at the time governance modes are selected. Exogenous and endogenous variables cannot be clearly distinguished in the model. Therefore, the model does not try to determine or predict governance modes, but rather to characterise them and associate them with certain traits. This is a clear limitation of this model.

Table 5.3 provides a short overview of how the variables behave in the four different governance modes. For each variable, the mean and standard deviation (S.D.) is indicated. For all dummy variables, the mean can be interpreted as the percentage of firms scoring a 1. Descriptive statistics show that hierarchical relationships go along with high values of the variables *location in China*, *working experience* and *dependency*. Markets relationships go along with high scores in the variables *time of negotiation* and *unit in HK*. All other variables do not seem to correlate with any governance mode. To analyse which variables are significantly correlated and have the strongest association with governance modes (highest β coefficient), an ordered logit model was applied, as the dependent variable is of ordinal scale.

It continues to be fairly common practice to analyse ordered choice variables using a linear regression model (LRM), because of the ease of interpretation. The advantages of using a LRM are the ability to use a simple estimator (ordinary least square (OLS)), the ease of interpretation and the small sample size required to build a model. But it also suffers from disadvantages, because (1) it requires a

Table 5.2.: Definition and descriptive statistics of variables used in the generalised ordered logit (GOL) model (1)

Variable	Unit	Mean	Std. Dev.	Min	Max
			$n = 206$		
Relation specific characteristics					
Location in China	dummy	0.55	0.50	0	1
Working experience	years	7.72	5.56	0	25
Dependency	%	52.77	30.98	2	100
Time for negotiation	hours	157.02	393.73	0	4368
Firm specific characteristics					
Predictability of supply/sales infos	Likert scale	2.36	0.83	1	5
Supply/sales unit operated in HK	dummy	0.71	0.46	0	1
Supply/sales unit family led	dummy	0.32	0.47	0	1
Supply/sales unit linked to externals	dummy	0.20	0.40	0	1

Source: Calculation based on own survey conducted in DFG SPP 1233 [2007]

lot of assumptions (linear correlation, expected value of errors $E[\varepsilon_i|x_{ik}] = 0$, homoscedasticity $V[\varepsilon_i] = \sigma^2$, no autocorrelation of errors ε_i, normal distributed errors $\varepsilon_i \sim N(0, \sigma^2)$, no exact multicollinearity of independent variables), (2) it assumes a constant change of variables across all categories of the independent variable, which is often unrealistic and (3) predictions $\hat{\pi}_i$ are often outside the boundary ($\hat{\pi}_i < 0$ or $\hat{\pi}_i > 1$). If this assumption is false, then LRM results may be biased and can be severely misleading (LONG and FREESE 2006:183–222). In contrast, the logit models avoid some of the problems of the LRM. Firstly, they work without the numerous assumptions in the LRM. They do not require a linear correlation, they have built-in homoscedasticity and errors ε_i do not need to be normally distributed, but they are fixed to $V[\varepsilon_i] = \pi^2/3$. Only the absence of multicollinearity should be considered. Secondly, in the logit model, no constant predictors are assumed anymore. Thirdly, the predicted probability always ranges from 0 to 1. However, logit models also have disadvantages compared to LRM. Instead of the easy-to-calculate OLS estimator, logits require the user to work with the more difficult maximum likelihood (ML) estimator. Additionally, the interpretation of logit models is more difficult (STEENBERGEN 2008:1–8).

Table 5.3.: Definition and descriptive statistics of variables used in GOL (1) according to their governance modes

Variable	H ($n = 59$)		EC ($n = 20$)		NEC ($n = 45$)		M ($n = 82$)	
	Mean	S.D.	Mean	S.D.	Mean	S.D.	Mean	S.D.
Relation specific characteristics								
Location in China	0.97	0.18	0.85	0.37	0.42	0.50	0.26	0.44
Working experience	11	6.11	7	4.91	6	4.00	7	5.21
Dependency	79	27.62	66	25.83	47	26.60	34	20.41
Time for negotiation	39	99.90	125	193.99	229	656.23	210	355.54
Firm specific characteristics								
Predictability	2.42	0.99	2.18	0.77	2.36	0.81	2.36	0.75
Supply/sales unit in HK	0.44	0.50	0.95	0.22	0.76	0.43	0.82	0.39
Supply/sales unit family led	0.31	0.46	0.35	0.49	0.22	0.42	0.37	0.48
Supply/sales unit linked	0.10	0.30	0.40	0.50	0.20	0.40	0.22	0.42

Source: Calculation based on own survey conducted in DFG SPP 1233 [2007]

In order to establish which variables are significantly associated with the 4-level ordinal scaled governance modes of HK firms, an ordered logit model (OLM) was chosen. This was preferred to an ordered probit model because model fit statistics are easier to calculate, which ensures quality measurement in the model design. The distributional assumptions for an OLM form a standard logistic function, which means:

- [1]$F(\varepsilon_i) = \Lambda(\varepsilon_i)$,

- [2]Thus $\Pr(y_i = m) = \Lambda(\tau_m - \mathbf{x_i}'\beta) - \Lambda(\tau_{m-1} - \mathbf{x_i}'\beta)$ and

- $\varepsilon_i \sim N(0, \frac{\pi^2}{3})$.

---

[1]Λ standard logistic cumulative distribution function
[2]Pr = Probability, y_i = independent variable, m = Alternative m in the choice set of the y_i, τ_m = cut points for m, $\mathbf{x_i}'$ = vector of respondent attributes, β = coefficients

Two assumptions have to be proven. Firstly, the absence of multicollinearity should be tested. This is proven in Appendix B in Table B.1. The Pearson correlation coefficients are very low and insignificant. Only the variables *location in China* and *dependency* have a significant correlation coefficient. As the first variable is a dummy variable, this correlation must be analysed carefully, because it could be created by chance. Therefore, it has been decided to keep both variables in the set of predictors. Secondly, the OLM initially assumes that the effect of predictors remains constant across the categories of the response variable. This is known as the parallel regression assumption (PRA). When this assumption fails, the estimators become biased and inconsistent. It is easy to see why. Imagine that a predictor has a positive effect for some categories of y and a negative effect for other categories. When the effect is constrained to become constant across all of the categories of y, as we have to assume under the PRA, then the net effect might well come out as zero. The PRA is a rather stringent assumption that is frequently violated (LONG and FREESE 2006:197–200). The Brant Test indicates whether the PRA is violated and which variables are responsible for the violation. For the ordered logit model designed above, the variables *unit is linked to outside supporters* and *unit is led by family members* violate the PRA. As recommended by LONG and FREESE (2006:200), the generalised ordered logit model (GOL) solves the problem, because the PRA has been relaxed for the two variables, which allows them to vary across the categories of Y. The possibility of relaxing the PRA in a GOL is its advantage over LRM.

Table 5.4 shows the GOL (1) for the different governance modes for customer and producer relationships. The model clearly indicates that *location of partners* has the largest influence ($\beta = -1.8167^{***}$). If the location of the most important customer or producer is in China, firms opt significantly more frequently for a hierarchical form of relationship. Firms in HK opt for hierarchical relationships if they work within China, whereas they establish market relationships with international partners. This trend is significant. The unusually strong links between firms in HK and the PRD might be caused by the incomplete institutional framework in the PRD. In order so survive in a transforming economic environment, firms need to find a governance mode which mitigates the political hazards and copes with procedural risk while remaining competitive internationally (KRUG and HENDRISCHKE 2006b). When HK firms adapt to the difficult institutional setting in the PRD, it seems to be more appropriate and safer to maintain control of the PRD operations and set up self-controlled affiliates, instead of dealing with independent producers or customers and taking the risk of not knowing whether the partner is able to master the situation. But with the improvement of the institutional environment, cooperation and the forming of networks can be expected (WILLIAMSON 1998; KRUG and HENDRISCHKE 2006b). In contrast, HK firms' trade with partners in developed countries is already protected by international trade security regulations. It

Table 5.4.: GOL (1) for governance modes in customer producer relations of firms in HK

Variable	Mode	Coef.	S. E.	z	P > z	[95% Conf. Interval]	
Location in China		−1.8167	0.3533	−5.14	0.000***	−2.5092	−1.1243
Working experience		−0.0829	0.0293	−2.83	0.005***	−0.1403	−0.0254
Dependency		−0.0331	0.0058	−5.69	0.000***	−0.0445	−0.0217
Time for negotiation		0.0000	0.0004	−0.01	0.993	−0.0007	0.0007
Predictability		−0.2049	0.1781	−1.15	0.250	−0.5540	0.1442
Supply/sales unit in HK		0.6871	0.3481	1.97	0.048**	0.0048	1.3694
relaxed PRA for							
Supply/sales unit family led	> H	−0.1144	0.4428	−0.26	0.796	−0.9822	0.7534
	> EC	−0.5587	0.4225	−1.32	0.186	−1.3869	0.2694
	> NEC	0.2051	0.4047	0.51	0.612	−0.5881	0.9983
Supply/sales unit linked	> H	1.1090	0.5851	1.9	0.058*	−0.0378	2.2558
	> EC	−0.4158	0.4871	−0.85	0.393	−1.3706	0.5389
	> NEC	0.0401	0.4622	0.09	0.931	−0.8658	0.9460
Constant	> H	4.7859	0.7748	6.18	0.000	3.2673	6.3046
	> EC	4.3343	0.7568	5.73	0.000	2.8510	5.8176
	> NEC	2.4506	0.6991	3.51	0.000	1.0804	3.8209

Dep. Var.: Governance modes for customer and producer relations (1=Hierarchy (H), 2=Equity cooperation (EC), 3=Non-equity cooperation (NEC), 4=Market (M))

Log likelihood	= −189.3	Wald χ^2_{12}	= 96.4
Number of obs	= 206	Prob > χ^2	= 0.000
Correctly classified cases	= 63%	$R^2_{McFadden}$	= 0.284
		$R^2_{CraggUhler}$	= 0.561

Source: Calculation based on own survey conducted in DFG SPP 1233 [2007]

can therefore be cautiously concluded that the incomplete institutional environment is responsible for this decision.

Comparatively much less important, but nevertheless significant, are the *years of working experience* ($\beta = -0.0829$***). The longer the working experience, the more often transactions are embedded in hierarchies. As markets ease the switch of partners, these relationships naturally become shorter, but in the PRD, the location-specific institutions act cumulatively on that pattern. Relationships set up before China entered the WTO and CEPA were difficult to manage on a market basis. Firms which established contracts after 2001 could enjoy a much more stable situation in China, which has encouraged more firms to work within cooperation or sometimes even in market-like relationships. WANG and NICHOLAS (2007) also emphasised the recent trend towards cooperation in their PRD study. Firms seem to screen the institutional environment constantly in order to choose the appropriate governance mode. Moreover the *degree to which HK firms depend on their major customers or producers* is significantly associated with a certain governance mode ($\beta = -0.0331^{***}$). The higher the profits made with one partner, the more likely the relationship is to be organised in a hierarchical way. This means that if HK firms have set up their own subsidiary, then most of the production is run by that subsidiary. Using the market place means choosing the best partner and spreading risk, which usually results in a more diverse pool of customers and suppliers. Additionally, when *firms operate their sales or procurement unit in HK*, they have market-like relationships significantly more often ($\beta = 0.6871^{**}$). Other variables such as *time of negotiation, predictability of information related to producers or customers, a family led unit* and *having a unit linked to outside advisers* are not significantly connected with specific governance modes. There is an association between the *degree of dependency on partners, the work experience* and *certain governance modes*, but while both variables describe specifications of governance modes, they are not decision-making criteria for a mode. Only the *location of customers and producers* and the *location of decision-making* is previously known and can influence the decision. The exclusion of the influence of other variables leads to the conclusion that differences in governance of customer and producer relations can be explained by locations connected to different institutional settings.

In order to qualify the findings of the data, the model fit must be examined. In the context of logit models, there are three useful approaches towards assessing model fit: correct predictions, pseudo-R^2 and the likelihood ratio test. Based on the predicted probabilities, one can compare the actual selection of governance modes for each case to the predicted choice of governance modes calculated by the GOL. The percentage of correct predictions can be calculated. In the GOL (1), 63% of the cases are correctly classified, which is a sufficient measurement. Moreover, the pseudo-R^2 of McFadden (0.284) and Cragg-Uhler (Nagelkerke) (0.561) can be taken as adequate for model fit (LONG and FREESE 2006). The likelihood ratio test contrasts the given model with an empty model, e. g. a model where all coefficients are zero and which contains a constant only. If the empty model fits the data no worse than the substantive model, then one cannot reject the nill hypothesis that none of the predictors matter. When referred to a χ^2_{12}-distribution in GOL(1) $p = 0.000$, one can clearly reject the hypothesis that none of the predictors matter. The model fit statistics show that GOL(1) is well defined.

Within the model, it is possible to obtain a predicted probability by simulating a set of values on the covariates. When there are many predictors, however, it is sometimes helpful to keep most covariates constant and to assess the effect of only a few covariates. For example, one can assess the predicted probability across the range of values of one predictor while keeping all other predictors at their mean. In Table 5.5, predicted probabilities for different firm types are calculated. When partners are located in China (dummy = 1, while keeping all other variables on their mean), the predicted probability of being involved in a hierarchical relationship is 32%. The probability of having a market relationship is only 16%. Surprisingly, the probability for being in a NEC is 31%. This could be because of the improving institutional environment, which allows room for NEC. In contrast, having a partner abroad reduces the probability of having a hierarchical relationship to 7%, while increasing the probability of acting on market to 54%. This picture clearly shows the influence of location. The high score of NEC when having a partner in China raises the question of whether time matters for governance mode. The years of working experience became significant in the GOL(1). Moreover, it is of interest what modes firms prefer if they have set up their relationships to partners recently and what modes are most successful. When firms have little working experience with their partners, they are more likely to be organised in markets or NEC. In contrast, the likelihood of working in hierarchies increases when a long-term relationship exists. A good predictability of partner-related information is more likely when firms are organised in NEC or markets. A poor predictability is more likely to be observed in hierarchical relationships. Calculating the predicted probabilities for having little working experience and a good predictability simulates a company which has set up their relationships recently, but has a good predictability of information regarding the behaviour of partners. The probability of those firms acting on markets is 45% and in NEC 34%. This shows that successful predictability of change can be perfectly supported by young relationships on markets and in cooperation. In comparison, long relationships involving extensive working experience but poor predictability are likely to be organised in hierarchies (39%). Overall, this table shows that location matters and that younger relationships are differently organised without suffering from poor predictability when the right governance mode is selected.

It has been shown that the location matters for the characteristics of governance modes. HK firms act significantly differently when working with partners in China as opposed to those abroad. It could be proven that the years of working experience have an effect on governance modes. Besides the relation-specific characteristics, firm-specific characteristics might have an effect on governance modes as suggested by the transaction cost theory. To prove this, it seems worth separating the customer and producer samples. A concentration on the producer sample of HK firms implies the opportunity to discover the influence of structural variables. Additionally, it provides the chance to see whether changes in the institutional setting in China result in changes in governance modes, while focusing on the working experience with producers.

Table 5.5.: Predicted probabilities using GOL (1) for different firm types

Different Types	Hierarchy	EC	NEC	Market
Partner located in China	0.3167	0.2148	0.3111	0.1573
Partner located abroad	0.0701	0.0857	0.3097	0.5345
Little working experience	0.1223	0.1320	0.3625	0.3832
Long working experience	0.2735	0.2061	0.3335	0.1869
Good predictability	0.1348	0.1413	0.3668	0.3570
Poor predictability	0.2613	0.2028	0.3394	0.1966
Little working experience, good predictability	0.0954	0.1098	0.3441	0.4508
Long working experience, poor predictability	0.3927	0.2202	0.2691	0.1180

Source: Calculation based on own survey conducted in DFG SPP 1233 [2007]

5.2.2. Governance Modes, Firm-Specific Characteristics and Institutional Change: Logit Model

For a better understanding of how HK firms organise their cross border production activities with China, a second logit model was applied to shed light on the special situation between HK and China. This model hopefully clarifies how *changes* in the institutional environment effect governance modes. The model only concentrates on PRD producer-related variables. The set of variables differs from the set in GOL (1). A locational variable is generated by measuring whether the *location* within the PRD matters for certain governance modes (see Table 5.6). SZ and DG were the core cities (dummy: core PRD = 1) which were entered by foreign firms. Nowadays they are highly developed. In comparison, having a producer in the remote areas of the PRD might be associated with a different governance mode. Other relationship-specific characteristics, such as *working experience, time for negotiation* and *dependency*, go along with the definitions of the first model. Looking to firm-specific characteristics, *firms' predictability of producer-related information* reflects how firms can deal with the need for flexibility. Moreover, the literature specific to transaction costs discusses that firms investing more capital (for example innovative firms) tend to organise their relationships more hierarchically to protect business. The variable *innovativeness* consists of classes reflecting the importance of innovation activities of firms in product, process, organisation and marketing. A high score indicates a high degree of innovativeness, while a low score indicates that firms are engaged in little or no innovation activities. Furthermore, it will be tested whether the *position in the value chain* influences the governance modes. Therefore, the GOL (2) includes the percentage of firms' sales generated by the

Table 5.6.: Definition and descriptive statistics of variables used in the GOL (2)

Variable	Unit	Mean	Std. Dev.	Min	Max
			$n = 100$		
Relation specific characteristics					
Core PRD	dummy[a]	0.79	0.41	0	1
Working experience	years	8.89	5.88	0	25
Time for negotiation	hours	62.47	106.26	0	504
Dependency	%	70.46	29.39	10	100
Firm specific characteristics					
Predictability of producers	Likert scale[b]	2.31	0.92	1	4
Innovativeness	4 classes[c]	1.68	0.84	0	3
Position in VC	%[d]	33.33	36.27	0	100
Structural characteristics					
Foundation year	5 classes[e]	3.19	1.26	1	5
Sales growth	4 classes[f]	2.84	1.03	1	4
Employees	8 classes[g]	1.98	1.44	0	7

[a] 1=Shenzhen/Dongguan 0=remote PRD
[b] 1=very predictable to 5=not predictable
[c] 0=not innovative to 3=very innovative
[d] Share of ODM/OBM production
[e] ≤1978, 1979–1990, 1991–1996, 1997–2000, 2001+
[f] <0%, 0–2%, 2–10%, 10%+
[g] 0, 1–5, 6–10, 11–15, 16–25, 26–50, 51–100, 101–200

Source: Calculation based on own survey conducted in DFG SPP 1233 [2007]

production of ODM and OBM goods. Structural characteristics are included as control variables, such as the *year of foundation* to see whether age matters, the *number of employees* as an indicator of a firm's size, and the *growth rate of sales* as an indicator of a firm's growth. The growth rate of sales is strongly correlated with the size of sales, the growth rate in profits and the return on investments. Therefore, both variables are included. A variable indicating ownership has been excluded because 78% of all firms in the sample are HK-owned, which makes it difficult to analyse differences between the governance modes. The correlation matrix according to Pearson is presented in Appendix B in Table B.2. Correlation coefficients are low, which confirms that the set of variables is suitable for a logit analysis. Only

Table 5.7.: Definition and descriptive statistics of variables used in GOL (2) accord-
       ing to their governance mode

Variable	H (n = 56) Mean	S.D.	EC (n = 16) Mean	S.D.	NEC (n = 15) Mean	S.D.	M (n = 13) Mean	S.D.
Relation specific characteristics								
Core PRD	0.75	0.44	0.69	0.48	1.00	0.00	0.85	0.38
Working experience	11	6.16	8	5.32	6	3.97	5	3.53
Time negotiation	28	44.02	87	145.52	101	112.57	137	169.50
Dependency	80	26.76	71	28.17	63	24.63	37	20.16
Firm specific characteristics								
Predict. producers	2.41	0.99	2.31	0.70	2.07	0.70	2.15	1.07
Innovativeness	1.82	0.83	1.50	0.82	1.67	0.72	1.31	0.95
Position in VC	30	35.75	39	40.95	42	33.78	32	36.94
Structural characteristics								
Foundation year	3.04	1.25	3.44	1.15	3.40	1.30	3.31	1.44
Sales growth	2.93	1.04	2.75	1.06	2.87	0.64	2.54	1.33
Employees	2	1.66	2	1.22	1	0.72	1	0.73

Source: Calculation based on own survey conducted in DFG SPP 1233 [2007]

the variable *year of foundation* is correlated with the *working experience* and the *number of employees*. Younger firms naturally cannot have long relationships with their customers and producers. It also seems logical that younger firms have fewer employees. Despite these high correlations, the year of foundation is kept in the set of variables, as it functions as a control variable.

Table 5.7 provides a short overview of the variables classified according to their governance mode. The descriptive statistics give a first clue as to which variables might become significant in the logit model. The values of the variables *working experience, time of negotiation, dependency, predictability of producer-related information* and *innovativeness* seem to differ between the governance modes. Whether these variables also become significant can be seen by calculating the logit model. The Brant Test was applied to establish whether the PRA is violated. Again, the Brant Test suggested moving from an ordered logit to a GOL, because the variable *dependency* violates the PRA (see Appendix C in Table C.1). Because the sample is reduced to producer relationships only, the number of observations dropped down to 100. LONG (1997:54) proposed the following guidelines for the use of ML in small samples:

> "It is risky to use ML with samples smaller than 100, while samples over 500 seem adequate. ...if there are many parameters, more observations are needed ...A rule of at least 10 observations per parameter seems reasonable."

As the GOL (2) includes 10 parameters, a sample size of 100 seems to be sufficient, even if it is on the limit. Because the number of cases is very small, variables which are, in fact, significant may not be calculated as such, while other variables could become significant purely by chance. Despite these risks, it still seems appropriate to work with this logit model, as the number of cases is 100 and will not be increased in the future. The value of the empirical data is too high to rely only on descriptive statistics.

The calculated GOL (2) is presented in Table 5.8. Four variables became significant. The *location* within the PRD seems to affect the governance modes significantly ($\beta = 1.1408^*$). When producers of HK firms are located in the core PRD (SZ, DG), they are organised significantly more often in market or NEC relationships. This might be because firms in the core already have more experience in serving HK firms, since they were set up early. With increasing skills and quality of products, HK firms do not need to control producers, but can rather act on cooperation. Another reason for this could be that the market dynamic in SZ and DG is hardly affected by policy makers, whereas in the areas developed later, market institutions are not so well enforced and governmental officials still have a lot of power. This might reduce the willingness of HK firms to work in market relationships within the areas of the PRD developed later.

The variable *working experience* also became significant ($\beta = -0.1277^{**}$). The longer the relationship exists, the more likely relationships are to be organised in hierarchies. This, again, might be for the various reasons mentioned above. Besides working experience and location within the PRD, the *dependency* of producers became significant. For dependency, the PRA has been relaxed as suggested by the Brant test. Different groups of governance modes can then be compared. As the table shows, dependency significantly distinguishes EC, NEC and markets from hierarchies ($\beta = -0.0193^{**}$). Its effect on distinguishing hierarchies and EC from NEC and markets is even higher ($\beta = -0.0296^{***}$), while the effect increased again when comparing markets to the other three modes. The dependency significantly decreases when moving from hierarchies to markets. But the changes from markets to NEC are associated with the greatest change in dependency.

The variable *time of negotiation* became significant as well. The time in hours for HK firms from initiating an order until all details are cleared and the production process can start significantly increases when moving from hierarchies to markets

Table 5.8.: GOL (2) for governance modes in producer relations of HK firms

Variable	Mode	Coef.	S. E.	z	P > z	[95% Conf. Interval]	
Core PRD		1.1408	0.6009	1.90	0.058*	−0.0370	2.3187
Working experience		−0.1277	0.0605	−2.11	0.035**	−0.2463	−0.0091
Time negotiation		0.0035	0.0020	1.77	0.077*	−0.0004	0.0074
Innovativeness		−0.1491	0.2973	−0.50	0.616	−0.7318	0.4337
Predictability producer		−0.2707	0.2593	−1.04	0.297	−0.7789	0.2376
Position in VC		0.0086	0.0064	1.35	0.177	−0.0039	0.0212
Employees		−0.1725	0.2117	−0.81	0.415	−0.5875	0.2425
Foundation year		−0.0319	0.2482	−0.13	0.898	−0.5183	0.4545
Sales growth		−0.3573	0.2404	−1.49	0.137	−0.8284	0.1139
relaxed PRA for							
Dependency	> H	−0.0193	0.0093	−2.08	0.038**	−0.0376	−0.0011
	> EC	−0.0296	0.0108	−2.73	0.006***	−0.0508	−0.0084
	> NEC	−0.0594	0.0177	−3.36	0.001***	−0.0941	−0.0248
Constant	> H	3.1097	1.5962	1.95	0.051*	−0.0188	6.2381
	> EC	2.6693	1.5807	1.69	0.091*	−0.4288	5.7674
	> NEC	2.7008	1.6131	1.67	0.094*	−0.4609	5.8625

Dep. Var.: Governance modes for customer and producer relations (1=Hierarchy (H), 2=Equity co-operation (EC), 3=Non-equity cooperation (NEC), 4=Market (M))

Log likelihood = −88.8   Wald χ^2_{12} = 39.3
Number of obs = 100   Prob > χ^2 = 0.000
Correctly classified cases = 67% $R^2_{McFadden}$ = 0.240
        $R^2_{CraggUhler}$ = 0.474

Source: Calculation based on own survey conducted in DFG SPP 1233 [2007]

($\beta = 0.0035^*$). When working in hierarchies, firms indicated a mean of 28 hours (see Table 5.7) to clear order details with their branch in the PRD. When HK firms work with a JV partner in EC, they need 87 hours to clear details, because they have to start negotiation processes with their JV partner. In NEC, the time of negotiation comes to 101 hours, and in markets to 137 hours. In order to meet market flexibility, hierarchies are able to react fastest in dealing with orders. Production processes start approximately two days after orders are placed. The time of negotiation is a measurement of quantitative flexibility. There is a successive increase of time from hierarchies to markets. As expected, in-house production organisation reduces time required to place orders to the PRD producers, whereas negotiation with a cooperation partner or market participant takes more time.

Although the *predictability of producer-related information* is insignificant, the descriptive statistics show that firms setting up their own plant in the PRD have more difficulties predicting market changes. They compensate this low predictability with greater flexibility when orders come in. In contrast, firms acting on markets or NEC can better predict producer-related information, such as knowing where raw materials can be sourced quickly, but they lose more time in negotiation. Hierarchical and market-related forms of organisation are both competitive, but in different ways. Hierarchies suffer from low predictability, but are able to find quick solutions. Relationships based on markets and in NEC enjoy a better predictability, but need more time for negotiation.

Other variables did not affect the governance modes significantly, but it is still worth mentioning in which direction they influence the governance modes. The theory suggested that the more innovative firms are, the more likely they would be to protect their business activities in hierarchies. This can be confirmed by the model. Firms which indicated being *highly innovative* more often prefer their producer to be a subsidiary rather than a market supplier. In contrast, firms in a higher *value chain position* (more sales generated by ODM/OBM production) tend to act in market-like relationships (positive β). Innovativeness of firms does not necessarily correlate with more ODM/OBM production. The control variables are insignificant.

The effect of the years of working experience is of special interest, because it could provide evidence of changing patterns in governance modes caused by institutional change. Figure 5.1 presents the predicted probabilities for the years of working experience. One can see that the predicted probability for EC, NEC and hierarchies are similar in the first three years (about 30%). When relationships are longer-standing, the hierarchical form takes over as the most likely governance mode. But the figure also reveals that the probability of working in EC is above 20% when relationships are not older than ten years. Additionally, the probability of NEC is above 20% when relationships are not older than 7.5 years. The black vertical line at 8.9 years indicates the average length of working experience. Although hierarchical relationships are still the most likely, it can be seen that the likelihood of cooperations has gradually increased in recent years. Cooperation is based on more informal agreements, because cooperation requires trust in people, trust in reliability and trust in the commitment to cooperate. These relationships seem to have increased in importance recently. Reasons for this are not easy to define, because they lie in different parallel developments.

1. The electronics value chain has changed from a hierarchical to a modular value chain. This development began in the 1990s. A modular value chain is based on cooperative relationships between customers and producers. Partnerships which were set up 10 or more years ago might still be characterised by strong hierarchical relationships, assuming that a certain persistency goes along with the selected governance mode. The gradual increase in coopera-

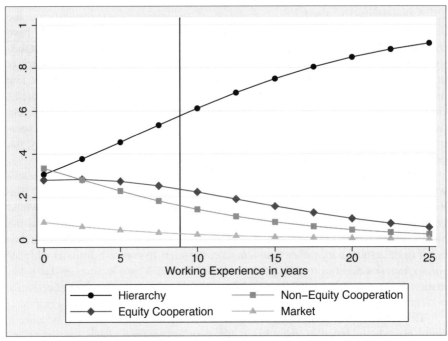

Figure 5.1.: Predicted probabilities for an increase in working experience

Source: Calculation based on own survey conducted in DFG SPP 1233 [2007]

tion for relationships characterised by an working experience of less than ten years can be seen as a sign of the shifting value chain pattern.

2. The shift from hierarchies to cooperation could also be a result of the improvement of the institutional environment in China. China began to open up in 1979. After ten years of attracting FDI and the simultaneous renewal and introduction of business and trading law to safeguard transactions in China and lower transactional costs, firms changed from hierarchical relationships to cooperation. They deemed safeguards to be sufficiently developed to change to a more market-oriented governance mode. Relying on EC or NEC was revealed as being most appropriate in the current institutional setting. Although pure market transactions are already taking place, they are less likely. Additionally, it must be considered that informal modes smooth cooperation. Firms do not necessarily profit from a market transaction, but rather benefit most when working in cooperation. Again, all the argumentation is based on the fact that governance modes are subject to a certain persistency. It can be argued that a change of governance modes generates transaction costs, which firms tend to avoid by staying with their chosen governance mode, at least for

a certain time. Therefore, relationships which were set up more than 10 years ago are more likely to be organised in hierarchies. If firms have invested in hierarchical relationships, for example building their own factory in the PRD, they will continue to produce most of their products in their own factory even if they would ordinarily opt for a different governance mode now. Unfortunately, there is a lack of empirical evidence to prove the persistency of governance modes.

3. The behaviour of the curves in Figure 5.1 can also be explained by general business practices. Firms first start to act on markets with their partners. After years of successful work with their partners, firms shift to cooperation. Firms trust in their partners' reliability and their commitment to cooperate. When firms cooperate for more than 10 years, they often buy their partner out and continue to work in hierarchies.

The effects of the three developments cannot be measured separately. It can be expected that all of them have a certain influence on the governance mode. The change of the electronics value chain and the improvement of the institutional setting in China in particular act in the same direction. But keeping in mind that younger relationships are more often organised using cooperation, and combined with the fact that in the core PRD, where institutions are even more stable, HK firms again work more often through cooperation and markets, one can cautiously conclude that the improving institutional environment is responsible to a high degree for the governance modes.

The model fit can be determined by the Wald test. The Wald test shows that one can clearly reject the hypothesis that none of the predictors matter ($\chi^2_{12} = 39.3, p = 0.000$). Moreover, the pseudo-$R^2$ of McFadden (0.23964) and Cragg-Uhler (Nagelkerke) (0.474) can be interpreted as sufficient for model quality. The model classifies 67% of cases correctly.

When looking at predicted probabilities for different types of firms, one can see that the likelihood of working in a hierarchy dominates the results. But there are still differences to discover which are worth mentioning when analysing the different characteristics of governance modes. Table 5.9 summarises the predicted probability of certain firm types which are associated with the four governance modes (sum = 100%). The table shows that firms' producer relationships in the *core* and *non-core PRD* are most likely to be organised in hierarchies. But the likelihood of being organised in informal governance modes (EC and NEC) is higher when a firm's producer is located in the core PRD. The core PRD seems to provide a better institutional environment and more highly qualified independent producers which HK firms can rely on. When looking at the *working experience*, firms with a long-term producer are most likely to be organised in hierarchies (75%). Firms with relationships to producers set up recently show a more diverse distribution.

Table 5.9.: Predicted probabilities using GOL (2) for different firm types

Different Types	Hierarchy	EC	NEC	Market
Core PRD	0.5183	0.2564	0.1899	0.0353
Non-Core PRD	0.7710	0.1439	0.0734	0.0116
Little working experience	0.3921	0.2813	0.2691	0.0576
Long working experience	0.7491	0.1560	0.0818	0.0130
Core PRC, little working experience	0.3367	0.2820	0.3093	0.0721
Non-Core PRD, little working experience	0.6136	0.2218	0.1404	0.0242
Core PRD, long working experience	0.7015	0.1810	0.1010	0.0165
Non-Core PRD, long working experience	0.8803	0.0789	0.0355	0.0053
Excellent predictability of producer infos	0.4895	0.2645	0.2066	0.0395
Poor predictability of producer information	0.7390	0.1615	0.0858	0.0137
Little working experience, short negotiation	0.4249	0.2776	0.2469	0.0507
Long working experience, short negotiation	0.7738	0.1424	0.0724	0.0114

Source: Calculation based on own survey conducted in DFG SPP 1233 [2007]

The likelihood of being organised in hierarchies is 39%, the likelihood of working in EC is 28%, and in NEC 27%. Reasons for that are the changing institutional environment and the shift in value chains or general business practices. Further explanations are given above. When combining the two variables, the predicted probabilities reveal that firms in the core PRD with little working experience with their customers are most likely to be organised in EC or NEC. In contrast, firms outside the core with long-term partners are most likely to have hierarchical relationships. The location, with a particular institutional environment, and the years of working experience seem to matter for the association with certain governance modes. Although the variable *predictability of producer-related information* is not significant in the GOL (2), its influence still becomes obvious when comparing the governance modes to producers of HK firms with excellent and poor predictability. If firms are able to predict changes related to producers very well, their probability of working in a hierarchical relationship is 49%, but the likelihood of working in EC (27%) and in NEC (21%) is also worth mentioning. In contrast, firms having a poor predictability are organised in hierarchies with a likelihood of 74%. This clearly shows that cooperative, more informal governance modes offer better conditions for predicting changes, probably because the pool of contacts from which information can be gained increases. In times of rapidly changing market conditions, decreased product life cycles and political uncertainty in China, a high predictability is a means of managing the need for flexibility. The changing behaviour can also be observed when comparing firms with the ability to keep the negotiation time short. Firms indicating a short negotiation time and little working experience have a much higher predicted probability of working on a cooperation basis than firms with more working experience. The ability to speed up negotiation processes is a necessary tool for enhancing competitiveness in times where high flexibility is required. Informal modes based on personal understanding and interaction increase the ability to predict how producers will perform and shorten the time required to conclude agreements.

Moreover, the logit model also provides a dynamic component for measuring effects of changes in the independent variables on the dependent variable. Table 5.10 indicates the average marginal effects of all dependent variables on the governance modes. The average marginal effect indicates how the probability of being organised in a governance mode increases/decreases in percentage points when an independent variable increases by one unit while keeping all other variables at their mean. The average marginal effect is calculated by increasing the independent variable x_i by one unit for each sample case and then calculating the differences in the predicted probabilities. When looking at dummy variables, the effect can be interpreted as discrete change when moving from 0 to 1. Taking into account the significant variables, one can see that a change of a major producer's location from the remote PRD to the core PRD decreases the probability of working in a hierarchy by 18.9 percentage points (pp), while it increases the probability of working in EC, NEC and markets. An additional year of working experience increases the predictability of working in hierarchies by 2pp. One can multiply the effect by 10 to get change over 10 years. 10 more years of working experience decreases the likelihood of being organised in EC by 4.4pp, in NEC by 6.9pp and in markets by 9.8pp. When analysing the average marginal effects of the time of negotiation in hours, the table shows that two additional days of negotiation (48 hours) decreases the probability of working in hierarchies by 28.8pp and increases the probability of working in the other three modes. It is noticeable that something which has a positive or negative effect on hierarchies has the opposite effect on the other three modes, except for dependency. When dependency increases by 10%, the likelihood of working in hierarchies (3.2pp), EC (0.7pp) and NEC (0.7pp) also increases, whereas the likelihood for markets decreases by 4.5pp.

By way of critical comment, although the logit models provide insights into the interrelation of different variables and the selected governance modes, the ex-

Table 5.10.: Average marginal effects in the GOL (2)

	Unit	Average marginal effects dy/dx			
		Hierarchy	EC	NEC	Market
Relation specific characteristics					
Core PRD	dummy[a]	−0.1885	0.0498	0.0651	0.0736
Working experience	years	0.0211	−0.0044	−0.0069	−0.0098
Time for negotiation	hours	−0.0006	0.0001	0.0002	0.0003
Dependency	%	0.0032	0.0007	0.0007	−0.0045
Firm specific characteristics					
Predictability of producers	Likert scale	0.0447	−0.0093	−0.0146	−0.0207
Innovativeness	4 classes	0.0246	−0.0051	−0.0081	−0.0114
Position in VC	%	−0.0014	0.0003	0.0005	0.0007
Structural characteristics					
Foundation year	5 classes	0.0053	−0.0011	−0.0017	−0.0024
Sales growth	4 classes	0.0588	−0.0123	−0.0193	−0.0273
Employees	8 classes	0.0285	−0.0059	−0.0093	−0.0132

[a]dy/dx is for average discrete change of dummy variable from 0 to 1

Source: Calculation based on own survey conducted in DFG SPP 1233 [2007]

act coefficient, the predicted probabilities and the marginal effects are somewhat superficial, because they suggest a concreteness of the measurement which cannot be found in reality with such a small sample. The logit model in general is very vulnerable to changes in the variables. An attempt was made to affect changes in the models to see whether the significance of variables proves a certain resistance to changes. The decision was made to present models with robust significant variables to ensure quality in spite of the small sample size. It is clear that a larger number of cases would improve model quality, but in the absence of more cases, it nevertheless seems worth presenting the results of the logit models. However, necessary caution must be taken in interpreting results.

In the first logit model, it could be proven that HK firms organise relationships to customers and producers differently according to location. It has been argued that locational effects can be interpreted as institutional effects. Institutional effects are still being heavily discussed in the academic literature, but only on a theoretical basis. Empirical evidence is seldom provided. This work has attempted to close this gap. The primary data effectively show that locational and therefore institutional factors indicate the highest coefficient and are most influencing. The second model focussed on the producer relations of HK firms. It could be proven that firms which have set up relationships to producers recently are more likely to work in cooperation. There are strong arguments for the fact that the improvement of institutions is responsible for that, but other factors could not be entirely excluded. For more reliable results on this issue, additional research is needed.

5.3. SUMMARY AND DISCUSSION OF FINDINGS IN SECTION 5

This chapter has shown which governance modes HK and PRD firms preferentially use to organise their customer producer networks. Hypothesis C1—*HK firms govern their relationships to firms in China differently from relationships to firms abroad*—can be confirmed. Relationships to Western firms are more likely to be organised via markets, whereas relationships to firms in the PRD are mainly organised in hierarchies. Within the PRD, firms do not tend to have different relationships to customers and producers.

Additionally, it was tested whether the location of customers and producers of HK firms significantly influences the selection of governance modes. Locations in the Western world and locations in the PRD are associated with different levels of business uncertainty due to the distinct institutional settings. A logit model could show that location is the most important factor for determining the selection of governance modes. As the global value chain approach does not suggest a difference in customer and producer relations per se, it can be concluded that the institutional setting is responsible for a large portion of differences in governance modes of HK firms. Hypothesis C2—*differences in applied governance modes of HK firms can be explained by the institutional environment*—can be confirmed.

A second logit model tried to indicate the influences of the institutional change on the governance mode to HK producers. It was tested whether the length of working experience significantly affects the choice of governance modes. HK firms which have set up their relationship to their most important producer recently opted for cooperation or market relationships significantly more often than firms which set up their relationships ten years ago. This shows a crucial change in governance modes. The conclusion that this is due to an improving institutional setting is based on the assumption that governance modes are subject to a certain persistency over time. Institutional change might also not be the only reason. The shift of the electronics value chain from a hierarchical to a modular chain, and the fact that this is only due to general business practices, cannot be ignored. Therefore, Hypothesis C3—*the institutional environment is responsible for the shift from hierarchies to market-like relationships*—cannot be confirmed in this way. There are good reasons to do so, but as influences other than the institutional change cannot be excluded, it would be wrong to confirm this hypothesis. It can only be said that there is nothing which disproves the influence of the institutional environment, but a confirmation is not possible. Further research is needed to exclude other influences.

> **I** *Do HK firms organise their business relations to customers and producers in the PRD differently from their relations to customers and producers abroad? Is this a consequence of the different development stages of the institutional environment? Is there a trend toward hybrid forms of governance in China resulting from the transitional change?*

Research question **I** can now be answered. HK firms organise their relationships to customers and producers in the PRD differently from those to customers and producers abroad. This can be explained by the different development stages of the institutional environment. When business safeguards are ill-protected by laws or the enforcement of laws is poor, firms tend to use the protection afforded by hierarchies. Cooperation turns out to be an intermediate mode used when the institutional environment has improved, but business still lacks reliable protection. Markets are only used when the institutional environment is well developed. The discussion about transaction cost theory and governance modes driven by MENARD (2005), WILLIAMSON (2005) and PIES (2001) corresponds with the empirical findings. Firms adapt their governance modes to the institutional environment (discussed in Section 2.4). It could not be conclusively proven whether changes of governance modes over time are exclusively subject to the improvement of institutions, but certainly institutional change is partly responsible. In times of institutional change, firms tend to adjust their organisation in a gradual way. A time delay (WILLIAMSON 1998) could indeed be confirmed between the introduction of new laws and their enforceability. Firms only adjust when enforceability of laws is ensured.

Evaluating the empirical results gained in Chapter 4 and Chapter 5 provides some information to fill the gap of how the lower end of the electronics value chain is organised in the GPRD region. It quickly becomes obvious that HK CMs receive orders from Western LFs. LFs and CMs tend to develop a strong long-term cooperation and an equal partnership. Modules of construction are handed over from LF to CMs. In contrast, HK firms do not continue this strategy of having equal partners in the PRD. They tend to set up a 100%-owned production plant in the PRD, due to uncertainty in the institutional environment. Cross border production activities are highly protected by hierarchies and do not follow the trend of modularisation. There is a trend towards cooperation, but this does not dominate the chain. Firms in the PRD indicated organising producer relations in cooperation or markets. One can conclude that although the upper end of the value chain reflects the development of the electronics industry very accurately (GEREFFI et al. 2005; STURGEON

2002), this does not necessarily apply to the lower end due to locational influences. An especially poorly or well developed institutional environment can switch the form of governance within the chain. In conclusion, proponents of the value chain approach should consider the fact that a locational perspective is needed, because the chain concept is not compatible with the institutional environments. Chains can switch between modes. In the GPRD, modular chains switch to hierarchical modes in cross border operations, but within China, they are subject to restrictions.

6. ANALYSING INFORMALITY AND FLEXIBILITY IN BUSINESS RELATIONS IN THE GPRD

In order to study informality in business relations, one must go beyond the discussion of governance modes. A more intense analysis of the organisation and arrangement of certain parts of relationships provides a better insight into the use of informality and helps to answer research question **J**. One major point of analysis is the different use of informality in HK and the PRD, as well as the explanation for those differences. The different dimensions of informality outlined in Table 2.1 will be taken into account for the analysis. Moreover, this chapter aims to assess the effect of informal behaviour on flexibility. For the analysis, the focus was placed on the following interaction of firms with their customers and producers:

- contact and selection of customers and producers

- contractual arrangements

- enforcement mechanisms

The first Section 6.1 focuses on the governance modes and their impact on flexibility. After that, the discussion moves beyond the governance modes. According to the focal points outlined above, the contact and selection processes of customers and producers in the PRD will be analysed (see Section 6.2). After firms gain new customers or producers, they must deal with the issue of contractual arrangements. The different types of formal and informal arrangements will be analysed in Section 6.3. When examining the issue in greater detail, the link between informal agreements and increasing flexibility becomes especially apparent. Even if customers and producers are carefully selected and contractual arrangements are concluded, conflicts can still occur. Section 6.4 analyses the different formal and informal means of settling disputes. At the end of this chapter, the results will be contrasted with the hypotheses set up in the theoretical part of the work.

6.1. IMPACT OF GOVERNANCE MODES ON THE FLEXIBILITY OF FIRMS

It could be shown in Chapter 5 that a high speed of negotiation, indicating quantitative flexibility, is found in hierarchical realtionships more often than in markets. This section will study the qualitative aspect of flexibility between the different governance modes.

A wide variety of actions for dealing with sudden changes allows firms to benefit in times of high market volatility. Firms in HK and the PRD were asked to indicate how they deal with orders which exceed their production capacity. Firms reported on a Likert scale how often they apply certain actions (1=very often, 5=never). In the following analysis, only firms which indicated using an action "very often" or "often" are counted in order to give an overview of the relevance of certain actions. It is assumed that the decision of how to deal with the order in question depends on the mode of governance towards the producer. A cross table was created measuring the frequency of certain actions while working in a certain governance mode. Table 6.1 shows the firms in HK and the PRD according to their governance modes towards producers and the different ways of dealing with orders exceeding capacity (multiple answer set). As the HK questionnaire had been revised before implementation in the PRD, one category was added. In hierarchies, EC and NEC, HK firms preferred to let their most important producer manage the exceeding order under any governance mode. Firms tend to trust their most important producer in terms of having the flexibility to increase production capacity. The most important producer is not only the one which receives most of their orders, but also the one providing a specific buffer when orders exceed a firm's capacity. When HK firms work in market relationships with their most important producer, 23% of the firms replied that they subcontract to other companies when orders exceed their capacity. Only 15% use their most important producer. In markets, producer relationships do not seem to provide such a high flexibility. Firms in EC and NEC do not hesitate to subcontract to other companies as well (either acquainted or completely new firms). Because the dependency on their most important producer is lower than in hierarchies, they have a range of other producers they can rely on when orders exceed their capacity. When firms work in cooperation with their most important producer, they usually agree on a certain maximum production capacity to be provided by the producer. Therefore, HK firms working in cooperation more often rely on other firms as their first choice when flexibility is required (EC: 59%, NEC: 53%). Firms working in hierarchies rely less frequently on other producers and reject orders (18%). This shows that firms working outside hierarchies use more opportunities to deal with orders exceeding capacity than those using hierarchies. They are less flexible in quantitative ways (time of negotiation), but more flexible in qualitative ways (opportunities to react). When looking at PRD firms, they prefer to renegotiate when orders exceed their production capacity. It is assumed that HK firms follow the same strategy, but it was not explicitly asked in the questionnaire. PRD firms working in hierarchies frequently place orders exceeding capacity with their most important producer (53%). In comparison, firms in the PRD working in EC, NEC and markets use their most important producer and other companies more frequently. They have a more diverse pool of choices they can rely on. Firms in hierarchical-like relationships have to reject orders more often than firms building

Table 6.1.: Actions taken when orders exceed production capacity according to governance modes to producers

	Hierarchy	EC	NEC	Market
Firms in HK (n = 107)[a]	$n = 57$	$n = 17$	$n = 17$	$n = 13$
Place orders to the most important producer	67%	88%	77%	15%
Subcontract to other companies	18%	59%	53%	23%
Reject orders	18%	12%	24%	15%
Firms in the PRD (n = 215)[a]	$n = 20$	$n = 13$	$n = 42$	$n = 140$
Re-negotiate	68%	75%	66%	54%
Place orders to the most important producer	53%	17%	23%	16%
Subcontract to other companies	31%	8%	29%	20%
Reject orders	31%	8%	0%	4%

[a] Actions taken when production capacity is suddenly exceeded, measured on a Likert scale with
(1=very often to 5=never), but the table only provides the percentage of firms which scored for
1 (very often) or 2 (often) using a certain action.

Source: Calculation based on own survey conducted in DFG SPP 1233 [2007]

their producer networks on NEC and markets. The more market-like relationships
are, the more flexibly firms can react in terms of the available opportunities.

GOFFIN et al. (2006) found that a flexible producer is prepared to work on week-
ends to fulfill a short-term order. He is able to produce and deliver the parts quickly.
Inflexible producers are unable to fulfill these wishes. One can conclude that because most of the HK and PRD firms' producers buffer orders exceeding capacity,
they provide a certain degree of production flexibility. GOFFIN et al. (2006) also
found in their study that personal relationships are of great importance for a fric-
tionless partnership. This element certainly smoothes the relationship and makes
the producer willing to perform quickly.

In contrast, flexibility is also required when orders are reduced or cancelled.
The different actions taken by PRD firms according to their governance mode with
customers are shown in Table 6.2. One can see the share of firms which indicated
using these actions very often or often (multiple answers possible). As products
have already been produced, the relationship to customers provides clues as to how
firms deal with the problem. Unfortunately, the data are only available for PRD
firms. Most frequently, firms renegotiate with their customers in HK. This is most

Table 6.2.: Actions taken when orders are reduced according to governance modes with HK customers

	Hierarchy	EC	NEC	Market
Firms in the PRD (n = 88)[a]	*n = 16*	*n = 9*	*n = 6*	*n = 57*
Re-negotiate	56%	78%	83%	79%
Sell remaining products to other companies	7%	11%	17%	18%
Put products into stock	7%	11%	0%	29%
Cancellation charge incurred	29%	11%	33%	21%
Scrap products	7%	0%	0%	6%

[a] Actions taken when production capacity is suddenly reduced, measured on a Likert scale with (1=very often to 5=never), but the table only provides the percentage of firms which scored for 1 (very often) or 2 (often) using a certain action.

Source: Calculation based on own survey conducted in DFG SPP 1233 [2007]

frequently done by firms working in EC, NEC and markets. When working in hierarchies, 29% of the firms have contracts with an agreed cancellation charge. Products can be very specialised for the parent company and unsellable to others. Therefore, they sell very few products to other companies (7%). In comparison, market-like relationships allow a wider variety. 18% of the firms frequently sell remaining products to other firms, as they have a large range of customers. 29% of the firms put products into storage. Only a small proportion of firms scrap products. Because of the small sample size of firms in cooperation, those numbers are only of limited relevance.

It could be proven that a firm's qualitative and quantitative flexibility according to its governance mode is reciprocal. Firms working in hierarchies enjoy a higher quantitative flexibility, but a lower qualitative flexibility. When moving to cooperation and markets, the quantitative flexibility decreases while the qualitative flexibility increases. These results are obviously only based on the data collection in this survey, and for a generalisation, the focus must be on a larger set of variables measuring the flexibility. What could not be proven is that a cooperative, more informal governance mode is related to greater flexibility. Reasons for that can mainly be found in the small number of cases working in cooperation. As this study is of an explorative character and questionnaires were limited in length, only the relationship to the most important producer was covered. But in reality, firms work with more than one producer. The relationships to them and their role in keeping a firm flexible could not be studied, but this area deserves further attention nonetheless.

A first clue as to how firms strategically manage their producer network in order to maintain flexibility is provided by KRAJEWSKI et al. (2005). They studied Taiwanese electronics firms and followed five different producer relationships. They identified two types of producers. The Taiwanese firm had high value producers with restricted supply contracts and relatively infrequent schedule revisions. Furthermore, limits of changes in timing and size of orders were set. Additionally, the firm studied had producers characterised by flexible supply contracts and relatively frequent schedule revisions, which produces a high level of uncertainty for the producer. For a deeper understanding of how a flexible producer network is structured in the PRD, the author recommends further research which takes an entire producer network into account.

6.2. CONTACTING AND SELECTING CUSTOMERS AND PRODUCERS IN THE GPRD

In this section, it will be shown what relevance informal contacting and selecting processes have in comparison to formal ones. It will be established whether informality in recruitment processes serves as a substitute or complement for lacking formal institutions in HK and the PRD. The *contacting processes* will be analysed taking into account the customer relations of HK and PRD firms (see Section 6.2.1). The role of informality in *selection processes* will be assessed while focussing on HK firms' criteria for selecting producers for long-term cooperation (see Section 6.2.2).

6.2.1. Contacting Processes

Referring back to the different *dimensions of informality* (see Table 2.1), *personalisation* as a crucial dimension of informality has been analysed in the contact and selection processes of customers and producers. A bidding competition or an internet search to recruit new customers and producers can be ranked as very impersonal, an exhibition as moderately personal, while the recruitment within business and private networks using their guanxi enables managers to recruit on a personal basis (see Section 2.5). When managers can rely on their guanxi networks, they profit from the intangible, transferable and reciprocal character of the network. Firms thus save ex ante transaction costs for a screening process of their environment for new customers and producers, the collection of information and the establishment of the first contact. When managers can revert to networks, they enlarge their pool of choices for recruiting new customers and producers. This increases the qualitative flexibility. Moreover, under volatile industry and market conditions, time is essential for a firm's entry to market (quantitative flexibility). Personal networks take time to establish, but they are always ready to use once they are there. It is a quicker way to activate new customers and producers at short notice (MILLINGTON et al. 2006).

In Table 6.3, the importance to HK firms of seven different contact channels for gaining new customers is evaluated. Contact channels range from very formal (bidding competition) to very informal (private contacts). Firstly, HK firms assessed their importance on a Likert scale (1=very important to 5=not important). The table shows that recommendations by business partners (5) are the most important (scale: 2.4) for HK firms in gaining new customers. This requires a network of trustworthy business partners to receive recommendations. Whereas business contacts seem to be of importance, private networks (7) are not anymore (scale: 4.3). This confirms the change in the nature of guanxi from a private to a functional network (see Section 2.5.1). This functional guanxi network is driven by personal connections and profit orientation. Personal connections do not necessarily involve sympathy as in the case of private contacts. Therefore, business connections are associated with moderate informality, whereas private contacts are associated with a high degree of informality. Sales agents (4) are the second most important (scale: 2.6). Sales agents are responsible for contacting and keeping in touch with customers. The personal contact is important, as this creates an atmosphere of trust. Nevertheless, personal assets are of less importance than in business networks. However, sales agents account for moderate informality. In contrast, gaining customers at exhibitions (3) is more impersonal and gaining them on the internet (2) or through bidding competitions (1), which is at a very impersonal level and therefore very formal. Impersonal means, such as gaining customers at exhibitions and on the internet, are the third most important (2.7). Formal bidding processes, the contacting of former customers and private contacts are of minor importance. In conclusion, the most important contacting channels are those characterised by a medium degree of informality. HK firms still prefer contacting channels with a medium degree of informality, although the institutional environment would allow them to rely on formal means only. Apparently, a certain degree of informality is conducive to business success. HK firms proved that contacting channels do not formalise completely, but that informal modes retain their importance. This was expected by the outlined theory as well (ZENGER et al. 2002). Formal institutions do not necessarily go along with formal interactions, but provide room for the use of informal as well as formal actions.

The table also shows how the different contact channels are correlated with each other. Spearman-Rho has been chosen as the correlation coefficient because the importance of contact channels is measured on an ordinal scale. The table shows that HK firms which ranked bidding competition (formal mode) high/low also ranked private contacts, former business relations and business partners (informal modes) significantly higher/lower. This can be explained from two perspectives. Firstly,

Table 6.3.: Correlation analysis of formal and informal contacting channels of HK firms to customers

	Impor-tance^a	(1) very formal	(2)	(3)	(4)	(5)	(6)	(7) very informal
					$n = 104$ ⟺			
(1) Bidding competition^b	4.0	1	0.092	0.028	0.056	0.195**	0.369***	0.308***
(2) Internet	2.7		1	0.338***	0.042	0.130	0.092	0.152
(3) Exhibitions and fairs	2.7			1	0.150	-0.190	-0.002	0.225**
(4) Sales agents	2.6				1	0.142	-0.020	0.192
(5) Recommendation by business partners	2.4					1	0.201**	0.271***
(6) Former business relationship	3.7						1	0.516***
(7) Private contacts (e.g. family ties)	4.3							1

^a Importance is measured on a Likert scale with 1=very important to 5=not important, mean value indicated
^b Spearman-Rho, significance level is indicated by the number of stars

Source: Calculation based on own survey conducted in DFG SPP 1233 [2007]

Table 6.4.: Importance and frequency of contacting channels to customers comparing HK and PRD firms

Customer contacting[a]	HK firms[b] $n = 104$	PRD firms[b] $n = 220$
Bidding competition	20%	29%
Active Searching (exhibitions, sales agents, internet)	52%	65%
Business Contacts (former workers, recommendation)	37%	61%
Private contacts (e. g. family ties)	10%	34%

[a] Importance is measured on a Likert scale with 1=very important to 5=not important, the share of
firms scored for 1=very important and 2=important is indicated
[b] Customers HK:PRD $\chi^2 = 56.83, p = 0.000^{***}$

Source: Calculation based on own survey conducted in DFG SPP 1233 [2007]

firms combine very formal with very informal modes in their contacting strategy
(high importance of both), because relying either on formal or on informal modes
alone does not work effectively. Private contacts are no longer the only way to
organise business, but they complement nearly all channels to smooth relations.
Secondly, when firms ranked both categories as being of low importance (as most of
the firms did), they experienced that methods of moderate informality work best. To
summarise, private contacts are still important, but they do not sufficiently serve the
HK firms on their own. Firms in HK tend to combine informal and formal modes
or rely on moderate modes of informality to recruit customers most effectively.
Although they could rely on formal contacting modes alone, the combination with
informal modes seems to be more promising to succeed in business. In terms of
contacting, informality is an opportunity rather than a necessity for HK firms.

Table 6.4 compares the results of HK firms with the recruitment channels of
PRD firms. Categories were summarised to make a comparison possible. Again,
the degree of informality increased from the category "bidding competition" to the
category "private contacts". The table shows the share of firms in each location
which ranked the different channels as being very important or important. One can
see that the share of HK firms is lower than the PRD firms' share over all categories.
In contrast to PRD firms, HK firms used the entire spectrum of the Likert scale (1–
5) to rank the importance of different recruitment channels. PRD firms are retained
to rank negative values.

In both locations, active searching (exhibitions, sales agents, internet) is most
frequently indicated as being important, closely followed by business contacts.
Whereas both categories were ranked of similar frequency in the PRD (65% and

61%), active searching is more frequently deemed important in HK than business contacts (52% and 37%). In contrast, PRD firms placed private contacts on the third position, whereas HK firms preferred bidding competition. Although HK firms use informal elements to contact customers, PRD firms still rely more on them than HK firms do. Private contacts are used by one third of the firms in the PRD, whereas they are only used by 10% of the HK firms. This might be due to the location of customers. HK firms serve a global customer base, but 40% of the PRD firms' customers are located in HK, the other 60% divided in about equal shares between Chinese and international customers. PRD firms may use more informal modes for contacting customers in China and sometimes in HK, whereas they certainly apply more formal modes when they contact international customers. Testing the distribution of customers within the two categories for differences with the χ^2-test shows differences to be significant ($\chi^2 = 56.83, p = 0.000^{***}$). To summarise, for contacting customers, PRD firms more frequently deemed informal modes to be important than HK firms, because customers of PRD firms are more often located in China or HK, where personal, more informal modes succeed in gaining business. HK firms more frequently rated formal or mixed modes as being of greater importance, whereas private contacts do not have such a high importance for recruiting customers as they used to.

6.2.2. Selection Processes

It is expected that contacting of customers and producers is done on a continual basis. A selection occurs if potential customers and producers fit the firm's profile. Selection processes are of special interest when firms look for new cooperation partners. Section 5.2 outlined that cooperation with PRD producers has become more important for HK firms in recent years. Therefore, the selection criteria for producers in cooperation have been analysed. Selection criteria can range from very formal, without any involvement of personal assets, to very informal, which qualifies producers only because of personal relations (informality dimension: personalisation). Expertise (1) can be seen as an objective measurement of skills and is therefore very impersonal. A good reputation (2) is based on a pool of subjective measurements and is therefore deemed to be impersonal, but not as impersonal as expertise. Positive experiences in previous business dealings (3) is based on a firm's own subjective measurement of the producer's quality. It is linked to moderate personalisation. In contrast, selection criteria are more personal when they qualify producers to get along well with local workers, suppliers (4) and public officials (5). When a producer is qualified because of a private connection (6), this counts as very personal and therefore informal. In this study, it will be analysed how HK firms have selected producers for long-term relationships in EC and NEC. The differentiation between EC and NEC provides interesting results about firms' preferences within those relationships. When looking at EC, Table 6.5 shows that positive experience in previous business dealings is of the greatest importance (scale: 1.94), followed by a good reputation of producers (scale: 2.12) and the existence of personal relationships (scale: 2.18). HK firms' selection criteria for producers in EC are based on subjective quality measurements and personal relationships. Informal and moderately informal criteria are mainly applied. However, when turning to NEC, expertise as an objective means of measuring the quality of producers is of the greatest importance, along with a good reputation (scale: 2.00). The third most important factor is positive experience in previous business dealings (scale: 2.12). Whereas HK firms apparently need to smooth relationships to potential JV partners by informal modes, HK firms working in NEC set their priority on measurements of quality.

A look at the correlation matrix provides even more insights. When HK firms work in EC, the categories of good reputation and positive experiences correlate strongly with the fact that partners are important for getting along with workers, suppliers and public officials. EC still seems to be important for HK firms to "get things done in the PRD". Expertise of partners is only important at the second stage. Apparently, HK firms are responsible for the entire production process, whereas Chinese partners deal with surrounding parties (workers, suppliers, public officials). However, this stands in contrast to NEC. Expertise of partners is of the greatest importance. Looking at the correlation matrix, one can see that expertise is, in fact, negatively correlated with partners getting along well with workers, suppliers and public officials. It is noticeable that the quality measurements (more formal) correlate significantly with personal relationships (most informal measurements). Although modes of quality are very important, they are combined with the most informal modes in NEC. NEC seems to be a partnership relying on high quality productions, but it does not neglect a certain level of personal assets. Apparently, a successful partnership in NEC is characterised by expertise and a good reputation of producers in combination with mutual understanding. In contrast, EC follows another strategy. It is mainly applied to running the business on a functional level. This was the way in which HK firms were forced to work at the beginning of the Chinese opening policy. The way EC is seen by HK firms does not seem to have changed over recent years.

In this section, it could be proven firstly that HK firms still use informal modes to recruit and select partners, even though the institutional environment would allow a reliance on formal modes alone. This confirms again that a certain degree of informalisation smooths business relations. Secondly, it could be shown that firms tend to combine formal and informal modes in order to succeed. Formal and informal modes are not mutually exclusive, but rather complement each other. Thirdly, it has been established that PRD firms still rely more frequently on informal modes, or at least assess their importance as being greater. This is certainly due to the fact that

Table 6.5.: Correlation analysis of formal and informal selection criteria of HK firms for producers in EC and NEC

Equity Cooperation, $n = 17$	Importance[a]	(1)	(2)	(3)	(4)	(5)	(6)
(1) Expertise[b]	2.24	1	0.590**	0.471	0.133	0.255	0.249
(2) Good reputation	2.12		1	0.761***	0.596**	0.626***	0.318
(3) Good experiences in previous business	1.94			1	0.503**	0.665***	0.391
(4) Get along with local workers/suppliers well	3.24				1	0.748***	0.475
(5) Get along with public officials well	3.47					1	0.520**
(6) Existence of personal relationships	2.18						1

Non-Equity Cooperation, $n = 17$	Importance[a]	(1)	(2)	(3)	(4)	(5)	(6)
(1) Expertise	2.00	1	0.581**	0.464	−0.087	−0.020	0.168
(2) Good reputation	2.00		1	0.812***	0.337	0.372	0.627***
(3) Good experiences in previous business	2.12			1	0.323	0.117	0.484**
(4) Get along with local workers/suppliers well	3.18				1	0.333	0.066
(5) Get along with public officials well	4.24					1	0.137
(6) Existence of personal relationships	2.76						1

[a] Importance is measured on a Likert scale with 1=very important to 5=not important, mean value indicated
[b] Spearman-Rho, significance level is indicated by the number of stars

Source: Calculation based on own survey conducted in DFG SPP 1233 [2007]

the PRD is still in transition and formal interaction modes do not find their basic analogy in law and enforcement. MILLINGTON et al. (2006) also worked on the identification process of Chinese partners. Their findings also show that personal connections, and in particular guanxi, have helped UK- and US-invested manufacturers in China to select producers. They also observed, however, that guanxi nowadays consists of business rather than family and community connections.

Once firms have contacted and selected their potential customers and producers, they have to settle transaction-specific details. The next section provides evidence of how contractual arrangements are concluded.

6.3. CONTRACTUAL ARRANGEMENTS OF HONG KONG FIRMS

This section relates back to the theoretical discussion in Section 2.5.2, in which it was outlined how contractual arrangements can be characterised in a dynamic (framework agreement and renegotiation) and a static model (direct contracting). The distribution of firms within the different contractual models will be shown, thereby examining informal and formal modes of contracting. Two *dimensions of informality* are analysed: the *form* of agreements (written → verbal) and *precision* (precise → open-ended). The more open-ended an agreement is and the more details are agreed on verbally, the greater the informal character of a contractual arrangement is (see Table 2.1). The dynamic model will be compared to the static model. A distinction here has been made between the different governance modes. To gauge the precision of contracts, the points covered will be analysed. Furthermore, it will be examined how details are fixed (written or verbally). Finally, it will be established whether the dynamic model reduces the time of negotiation in general, and whether informal agreements have a particular effect.

This section concentrates exclusively on the analysis of the HK survey, as the complete analysis requires data which were not collected in the PRD survey. Even then, the number of cases considered in this section is limited. Some information was not provided by all firms and some cases seemed to be illogical. Therefore, the sample size was reduced.

The four governance modes considered in Section 3.2.2 can be characterised and distinguished further according to the contractual arrangements. Table 3.1 provides an overview of how hierarchies, EC and markets can use FAs to switch from a static to a dynamic model. FAs are concluded in advance. They fully or partly cover contractual details such as product specification, quantity and price, delivery time of products, exclusive production and conflict resolution. As production is especially time-sensitive in highly competitive markets, it is expected that FAs speed up the negotiation process when an actual order is placed, because certain details are already fixed in advance. The time from the placement of orders until the actual production starts should be kept as short as possible.

Section 3.2.2 outlined in detail how the contractual models are linked to the governance modes. When market participants conclude a FA, they no longer act with an anonymous partner on pure markets. If they agree to cooperate, market relationships change to NEC. Moreover, firms working in hierarchies or EC can have additional FA. It is assumed that firms sometimes have more than one affiliate or JV for production. Affiliates or JV might work quite independently from the parent company. Parent companies might take advantage from FA, allocating which jobs the different plants have to fulfil. Although this organisational structure is still characterised by hierarchies/EC, it is not as tightly organised and controlled, but is characterised by a loose form.

Table 6.6 shows the different governance modes divided into loose and tight forms, and indicates whether they work with (+) or without (–) FA. Additionally, it is shown how many of the HK firms surveyed apply the different modes to organise their relationship to their most important customer and producer. Looking at the responses of HK firms regarding their customer relationships, most of them opted for a NEC and a market relationship (36% and 61% respectively). The figures of producer relationships are more diverse. 30% of the HK firms work in tight hierarchies and another 16% work in loose hierarchies. Turning to EC, only 5% of the HK firms opted for a tight EC, whereas 14% work in loose EC concluding FAs in advance. This already shows how important it is to distinguish between firms working with and without FAs. 14% of the HK firms work in pure market relationships with their producers and another 22% have switched to NEC using FAs.

Furthermore, it has been assessed whether a FA with customers affects the time of negotiation when the actual order is placed. It is assumed that the dynamic contracting model provides advantages in negotation time compared to the static model. Negotiation time was measured in days. The mean value as well as the 95% confidence interval is given. When looking at customer relations, HK firms gain advantages from transforming a pure market relationship to a NEC. Instead of negotiating for 13 days before a final order is placed with the HK firms and the production process can start, it takes only 7 days in NEC. The exchange of expectations and the conclusion of a FA nearly halve the time of negotiation and speed up the production process in general. Unfortunately, the effect of FA in customer relations cannot be assessed in EC and hierarchies due to the low number of cases. NEC reduces a firm's negotiation time and therefore increases its quantitative flexibility in contract conclusion.

The effect of FAs on producer relations has also been analysed. In general, negotiation time increases when moving from hierarchies to markets. Comparing the effect of FAs within the different governance modes provides interesting insights. Passing on orders within hierarchies only takes 2 days. FAs have no effect on the time of negotiation. In EC, the result is the opposite of what was expected. In tight

Table 6.6.: Negotiation time of HK firms according to governance modes

| | FA | Customers | | | | Producer | | | |
| | | n | $\%$ | Days of negotiation | | n | $\%$ | Days of negotiation | |
				Mean	95% C. I.[a]			Mean	95% C. I.[a]
Tight Hierarchy	–	2	3	–	–	22	30	2	[1,3]
Loose Hierarchy	+	0	0	–	–	12	16	2	[1,3]
Tight Equity Cooperation	–	1	1	–	–	4	5	4	[3,10]
Loose Equity Cooperation	+	0	0	–	–	10	14	7	[0,14]
Non-Equity Cooperation	+	27	36	7	[3,10]	16	22	4	[2,7]
Pure Market	–	46	61	13	[7,18]	10	14	7	[2,13]
In total		76	100			74	100		

[a]Confidence Interval

Source: Calculation based on own survey conducted in DFG SPP 1233 [2007]

EC, the negotiation time is reduced to 4 days, while it takes 7 days when FAs are concluded. This might be due to the fact that in tight EC, one partner can govern and control the other partner's work without any agreement. Similiarities to hierarchical organisations are obvious. This unequal relationship reduces the time of negotiation in tight EC. Chinese-HK JVs that developed after the Chinese opening are characteristic for this group. Loose EC can be seen in relationships of HK firms with a set of equally stated Chinese JV partners. In that case, negotiation is more time-consuming. In contrast, when comparing NEC and markets, FAs in NEC almost halve the time of negotiation, from 7 to 4 days.

For both customer and producer relations, FAs positively affect market relationships, because a transformation to NEC halves the time required by HK firms in negotiation processes. What could not be proven is a positive effect of FAs on the negotiation time in hierarchies and EC. If any equity relationship exits, a tight control reduces the time of negotiation or maintains it at the same level. What must also be noted is the ratio of the negotiation time among the different governance modes. NEC enhances negotation time by 2 days in comparison to hierarchies. When considering the high costs of hierarchies, NEC seems to be a good alternative for future business.

The last paragraph revealed the relevance of the dynamic contracting model for firms in HK. Half of the relations to customers and producers are organised within FAs. This paragraph analyses the formality and informality of FAs. For customer and producer relations, it has been outlined which points are agreed in FAs (Precision of FA) and in which form (written, verbal or both) it has been done. Points which require attention when contracts are concluded are: product specification, quantity, price, delivery time, agreement on exclusivity, penalties for breach of contract. This derives from interviews with lawyers in the GPRD. If everything is precisely fixed, a formal contract has been concluded. If the points mentioned above are missing, then contracts are open-ended and therefore informal. The share of HK firms which have fixed all points in the FA are given in Table 6.7. If points are fixed in FAs, this could be in written, verbal or both forms. The share of HK firms using the specific forms is given. Using verbal FA refers to an informal agreement, while using a written form refers to a formal agreement.

When looking at customer and producer relations, the majority of HK firms fix important points in written form. In customer relations, an average of 51% of HK firms fixed all points in writing, while in producer relations, 46% of firms did so. Comparing customer and producer relations, it becomes clear that the number of written details in FAs is slightly higher with customers than with producers. A verbal agreement is only concluded by a very small number of firms (customers: 7%, producers: 5%). In producer relations, the combination of written and verbal agreements is preferred by about one quarter of the firms, while in customer relations the number drops to 21%. A summary of the share of firms which fix necessary details in one of the three forms (written, verbal, both) provides information about the precision of FAs. 79% of HK firms agree on the important points in customer relations before an order is placed. In producer relations, 76% of firms do so. Returning to the *form of agreements* as a dimension of informality, the share of firms having informal agreements is relatively low. Contracts tend to be written. However, the share of informal agreements with PRD producers is slightly higher than with international customers. When looking at the *precision* as the second dimension of

Table 6.7.: Preciseness and form of framework agreements of HK firms

	Customers (n = 27)						Producers (n = 38)					
	sum	Preciseness FA		Form of FA			sum	Preciseness FA		Form of FA		
		not fixed	fixed[a]	written	verbal	both		not fixed	fixed[a]	written	verbal	both
	%	%	%	%	%	%	%	%	%	%	%	%
Product specification	100	7	93	63	7	22	100	16	84	50	3	32
Quantity of products	100	22	78	48	7	22	100	13	87	55	0	32
Price of products	100	15	85	63	4	19	100	16	84	53	3	29
Delivery time	100	15	85	59	7	19	100	13	87	58	5	24
Agreement on exclusivity	100	37	63	33	7	22	100	47	53	32	5	16
Penalties for breach of contracts	100	33	67	41	7	19	100	42	58	26	16	16
Mean	100	21	79	51	7	21	100	24	76	46	5	25

[a]sum of shares written, oral and both

Source: Calculation based on own survey conducted in DFG SPP 1233 [2007]

informality studied in this section, the findings show that more than three quarters of the necessary details are fixed in one of the three forms. Only a small proportion of firms have imprecise and therefore informal contracts. HK firms mainly keep to formal modes in terms of form and precision of FAs, although there is a slight trend towards more informal agreements with PRD producers. HK firms apply informal modes when contacting and selecting customers and producers, but when it comes to contractual arrangements, HK firms mainly do not rely on trust and a hand shake, but prefer written agreements. When disputes occur, a written document is of more value than a person's word. What could not be proven in the quantitative data, but has been derived from the interviews, is that contracts are mainly written, although quality is not comparable with Western standards.

> "There are differences in the quality of contracts between China and the international environment. In the PRD, personal relationships are of great importance. Family ties and long-term business relationships are reasons for non-dedicated contracts. We have often experienced that Chinese firms have only rudimentary contracts, even for large projects."

> "It is not the mentality of Chinese people or society to have contracts. It really is a new thing. They are not really used to foreseeing problems. We always want to make sure that all the grey zones are removed. ...But the Chinese face these problems when they arise. They are changing, but it still takes time."

Often, a one page contract is concluded covering the main points, but it is not proofed by lawyers. Those contracts are very difficult to enforce by law in the case of disputes. For a better understanding, one has to keep in mind that points are fixed in a written form, but those written contracts are not comparable to what Westerners intuitively associate with written contracts.

> "Chinese firms only have a one page contract with each other, even if it is a big deal."

> "There can be a beautiful contract, but if it cannot be enforced, that work is worth nothing."

An important contribution of this work is to study the correlation of informality and flexibility. Firstly, it is of great interest to know whether firms can reduce the number of points they need to negotiate when orders are placed due to the conclusion of FAs in advance. Secondly, it must be investigated whether informal FAs are perhaps even of greater advantage than formal FAs with respect to the time of negotiation.

Taking the first point into account, firms with FAs were asked to provide information as to which points they renegotiate. It is expected that firms using the static contracting model without FAs must agree to all important points when orders are placed. In contrast, HK firms which have FA with customers indicated negotiating fewer points. 52% of the HK firms do not renegotiate product specifications. For quantity (41%) and price (48%) of products, almost half of the firms have no need for renegotiation. Delivery time is sensitive. Only 22% of firms can forecast delivery time and do not need any renegotiation. 19% of all firms indicated that they can start production immediately when orders are placed without any renegotiation process. The χ^2-test proved that FAs significantly reduce the number of points for renegotiation. Turning to producer relations, the share of HK firms which have to renegotiate details is higher than in customer relations. Only one third of firms do not consider renegotiating (product specification: 34%, quantity: 24%, price: 34%, delivery time: 18%). The share of HK firms without any renegotiation drops to 13%. Again the χ^2-test showed that FAs reduce the number of points for renegotiation significantly. Producer relations require more attention in renegotiation processes, because (1) FAs are less precise than in customer relations (see Table 6.7) and (2) although points are fixed, the standard quality of contracts is lower than in the Western business world. In contrast, contracts with international customers are mainly characterised by higher quality, meaning that points do not require further negotiation. It is expected that an increasing number of negotiation points is correlated with an extended negotiation process, which delays the production process. But overall, it could be proven that FAs reduce the points which need to be negotiated when orders are placed in customer and producer relations. This increases the flexibility of firms, because the negotiation process is expected to speed up accordingly.

It has been shown that FAs can reduce the number of negotiation points when orders are placed. This enables firms to respond to customers more quickly. In order to prove the correlation of informality and flexibility, evidence must be found of informal FAs leaving more leeway for firms to arrange the actual order, saving them extra time and reducing transaction costs. Informality in FAs was measured according to the number of details in contracts (product description, price, product volume, delivery time, exclusive production, conflict resolution) which are agreed on verbally rather than being fixed in written form. All details are taken to be equally important. A FA is defined as informal if most details are fixed verbally. It is assumed to be formal when most details are fixed in writing. Flexibility in contractual arrangements was measured according to the time in days required by firms to negotiate details before the production process can start.

Table 6.8 shows what effect informal FAs have on the time of negotiation in contrast to formal FAs, distinguishing between customers and producers. HK firms which have informal FAs with their producers indicated a shorter negotiation time on average than firms with formal FAs (3 days instead of 5 days). In this case, informal FAs seem to allow more leeway for final negotiation, which reduces negotiation time before production can start. In a rapidly changing business environment, the costs of drawing up a comprehensive contract are extremely high. It could be proven in Chapter 4 that PRD firms in particular, as component suppliers of HK firms, suffer from permanently changing market conditions and poor forecasting systems, which increases their need for flexibility. As the enforcing of contracts is difficult in the PRD due to the poor legal safeguards, HK firms tend to have either verbal contracts or poorly written contracts. Neither are up to legal standards.

Table 6.8.: Effects of formal and informal framework agreements on the negotiation time

| | Customers | | Producers | |
	n	Time of negotiation	n	Time of negotiation
Formal framework agreement	18	7 days	23	5 days
Informal framework agreement	9	6 days	15	3 days

Source: Calculation based on own survey conducted in DFG SPP 1233 [2007]

Two thirds of the firms revisited the FAs when orders were placed. However, this does not seem to negatively affect the negotiation time. Informal FAs even enable firms to agree more quickly on final contractual details. When producers are located within the PRD, informality in contractual arrangements works well in combination with spatial proximity and guanxi, as information can be verbally agreed on and revised. This speeds up the process of negotiation. The results were not significant, which is due to the small sample size. The correlation between informal agreements and quantitative flexibility in producer relations could be proven. When the need for flexibility is great because predictability of demand changes is low, informal agreements provide more leeway for negotiation and speed up the process. LYONS (1994) investigated firms in France and found that contracts between firms usually tend to be informalised when they work in rapidly changing business environments, because the costs of drawing up a comprehensive contract are too high.

Informal agreements with customers reduce the time required for negotiation when an actual order is placed from 7 days to 6 days. Within customer and producer relations, informal FAs reduce the negotiation time. But it seems to be more difficult to agree upon actual orders when the FA is informally organised in long-distance customer relationships, because the absolute negotiation time is longer. The effect of FAs is less strong than in producer relations. As forecasting systems of HK firms are better and the need for flexibility is lower, informal agreements are of advantage, but formal agreements can hold their own in terms of speed. As HK firms and international customers work in a complete institutional environment, transaction costs for concluding written contracts are lower. Additionally, this seems to be of advantage for the enforcement of contracts, as the legal system provides firms with sufficient enforcement mechanisms. Keeping in mind those results, the next section analyses the informal and formal enforcement mechanisms of contracts in customer and producer relations.

In this section the contractual arrangements between HK firms and their customers and producers are examined. It could be shown that FAs provide an opportunity to cover important points of negotiation in advance in order to save time when the order is placed, allowing the production process to start without delay. FAs can be applied in combination with each governance mode. Comparing the negotiation time of firms with and without FAs, it could be shown that firms in market relationships in particular benefit from FAs. If pure market relationships turn into NEC while concluding a FA, the negotiation time nearly halves. The effect of FAs in EC and hierarchies could not be proven as positive. Moreover, it has been shown that most of the FAs reach a formal stage, because they are concluded in written form and cover all necessary details. In customer relations, FAs are even more formal than in producer relations. Nevertheless, it could be proven for customers and producers of HK firms that informal FAs leave more leeway for negotiation and reduce the negotiation time compared to formal FAs. Informality in contractual arrangements enhances a firm's quantitative flexibility in terms of negotiation time. Although firms in the GPRD select partners carefully and try to optimise contractual arrangements, disputes can still occur and require the enforcement of contracts.

6.4. ENFORCEMENT MECHANISMS OF CONTRACTS IN THE GPRD

Although firms apply ex-ante safeguards to ensure the reliability of partners, disputes can still occur. Ex-post enforcement mechanisms are needed. Section 2.5.3 outlines the different ex-post enforcement mechanisms. Two *dimensions of informality* (see Table 2.1) are thereby touched upon: *enforcement* (tight $\rightarrow$ loose) and *power* (legal $\rightarrow$ social). Dispute resolution and contract enforcement can be very formally organised when firms opt for litigation. This implies legal power and a tight enforcement (close to the contract). However, firms can also decide on alternative dispute settlement. They might negotiate, mediate or arbitrate. Negotiation is the most informal mode of resolving disputes. It is driven by social power and may only loosely keep to the actual contract, but considers the situation as a whole. When negotiation does not lead to any satisfying result, firms can opt for mediation. Mediation is more formal than negotiation, but still provides more informal elements than arbitration. Arbitration can follow mediation or negotiation when no solution is reached. Arbitration is a direct alternative to litigation, but it is more confidential and ensures privacy. Its character tends to be less formal than litigation. In this section, it will firstly be outlined which means of dispute resolution are applied by HK and PRD firms, and secondly an attempt will be made to find determinants for the choice of different means.

It has been assumed that every dispute resolution process starts with negotiation, but of most interest is whether firms continue to apply formal enforcement mechanisms if negotiation fails. Therefore, data were collected about the variety of techniques applied by HK and PRD firms (see Table 6.9). Firms in HK and the PRD were asked which actions they had taken to settle a dispute within the last five

Table 6.9.: Enforcement mechanisms of contracts in HK and the PRD

	PRD survey[a] $n = 218$	HK survey $n = 100$
Negotiation only	62%	45%
Negotiation and mediation	27%	11%
Negotiation, mediation, arbitration and litigation	11%	44%
Total	100%	100%

[a]$\chi^2 = 113.0, p = 0.000$***

Source: Calculation based on own survey conducted in DFG SPP 1233 [2007]

years. 45% of the firms surveyed in HK indicated that they have only used negotiation to settle disputes. 11% have used negotiations followed by mediation and 44% reported also considering arbitration and litigation for dispute resolution. Nearly half of the firms in HK rely exclusively on informal dispute resolution processes. In contrast, firms surveyed in the PRD negotiate more often with their producers alone (62%). Another 27% move on to mediation when disputes occur. This share is higher than in HK, because in China, mediation is a common strategy used by courts. Mediation can be mandatory before a litigation process is able to start at all. Additionally, China has a long and rich philosophical history of mediation. The culture backed by confucianism holds that the best way to maintain harmony is to resolve a dispute through moral persuasion and agreement. A peaceful and respectful compromise should be reached. This also explains why only a few firms opt for arbitration and litigation (11%).

Returning to the contractual arrangements, it was established that contracts exist in a written form, but interviews with experts showed that this written form does not satisfy the legal requirements. Arbitration and litigation can only succeed when contracts are adapted to a legal form and enforcement of contracts is ensured by law. When contracts do not fulfil the legal requirements, they have no chance of being enforced by law. But if law and enforcement standards do not guarantee a fair judgment, firms do not invest transaction costs in adapting contracts to a legal form (chicken and egg problem).

Taking litigation processes into account, China is lacking laws adapted to the international standard and qualified education for judges to enforce them. For example in the past, military officers took positions as judges. Although the central government has improved legal standards and the education of judges, some firms fear not receiving a fair judgment in China. Transparency is lacking.

If contracts are properly formalised, firms in China also consider arbitration as an alternative. But arbitration in China also involves difficulties in application. In China there is a distinction between international and domestic arbitration. International arbitration is only allowed to be handled by the China International Economic and Trade Arbitration Commission (CIETAC) or by the China Maritime Arbitration Commission (CMAC). Foreign arbitrators are not allowed. As more international firms enter China, the arbitration commissions' workload has increased enormously. This is the *first* reason why arbitration is less attractive for firms (CHANG and HILMER 2006). Additionally, Chinese arbitration law is lacking in quality. CHANG and HILMER (2006) cited that "...there have been many uncertainties about the Arbitration Law and this is why China's People's Supreme Court addressed a number of judicial interpretations over the last years in an attempt ...to fill the gaps of the legislation". For the international harmonisation and unification of the law of international trade, the UNITED NATIONS COMMISSION ON INTERNATIONAL TRADE LAW (UNCITRAL) (n. d.) has been established. The UNICTRAL Arbitration Rules deal with procedural guidelines for parties in a commercial relationship which agree to accept arbitration. The Chinese Arbitration Law 1994 and the UNCITRAL Model Law have a number of differences, which is the *second* reason why arbitration in China is not very popular, as it does not meet international standards. CHANG and HILMER (2006) call for an urgently needed improvement and reform of the Arbitration Law 1994, as China's economy is growing and is becoming more international, leading to many international disputes that need to be settled through a rapid and efficient arbitration procedure. They suggest allowing foreign lawyers to practice law in China and revising the pricing policy. The Chinese government regulates arbitrators' fees, which do not comply with international standards. This gives rise to concern about the quality of arbitrators.

Although improvements have been seen in recent years, firms have not switched immediately to formal modes of dispute resolution. As WILLIAMSON (2000:597) outlined, there is only a slow and gradual change from the use of informal to formal means (see Figure 2.5). A change in a firm's behaviour takes time. Firms prefer to resolve conflicts informally among themselves. Even if the institutional environment continues to improve, it will take some time before Chinese firms can take advantage of the new techniques. Moreover, according to the WORLDBANK (n. d.)'s indicators "Doing Business", the enforcement of contracts through legal action takes 406 days (2008) without any change in the last 4 years, whereas it takes only 211 days in HK. It is expected that firms can greatly reduce this time using more informal means of dispute resolution.

A frequently used alternative to arbitration in China is mediation. Because arbitration and litigation need to be more institutionalised in China, firms in the PRD indicated mainly using negotiation and mediation techniques to settle disputes.

In contrast, HK firms firstly work under a complete set of laws protecting their trading and manufacturing activities accompanied by a functional court system and qualified judges, and secondly they enjoy a world class arbitration law and system, which gives them leeway to decide on the most appropriate alternative for settling disputes. HK has an excellent Arbitration Commission Centre, which not only has highly educated arbitrators, but also promotes arbitration among the firms. The applied arbitration rules of the centre are attractive for firms. Moreover, the HK government has established an Arbitration Law adapted to the UNCITRAL Model Law. Additionally, mediation is a hot topic of discussion in HK right now (HILMER 2007). The HK government wants firms to have the opportunity to mediate effectively, rather than going directly to arbitration. For this reason, HK has established a mediation centre. The use of mediation is a more cost-effective means of resolving disputes. HK hopes to establish itself as a hub for legal services. Two HK universities offer a masters degree programme in Arbitration and Dispute Resolution. As HK is a hub for international trade, dispute resolution must be fair, quick and cost-effective. Today, companies and enterprises doing business in mainland China and other Asian jurisdictions increasingly use HK as an arbitration and mediation hub (SANDBORG 2007). Providing firms with a set of alternative means for dispute resolution accompanied by the appropriate infrastructure increases their variety of choice and therefore their flexibility. The HK data show that a good legal system encourages more firms to find the appropriate dispute resolution process. Although HK firms can use courts, most firms still prefer negotiation as an informal mode of solving disputes, because in many cases, this is more appropriate. Negotiation is time- and cost-saving in comparison to litigation. Negotiation relies on the will to reach an agreement confidentially. This increases the probability of abiding by the terms of the agreement later on, even if it is not legally binding. But if negotiation fails, at least half of the firms surveyed in HK move on and use the entire spectrum of dispute resolution mechanisms to solve disputes. They tend to use formal and informal modes of dispute resolution at the same time. This makes them flexible while providing the chance to choose the appropriate mode of resolving disputes.

The distribution of firms in HK over the three categories is significantly different from the distribution of firms in the PRD ($\chi^2 = 113.0, p = 0.000^{***}$). PRD firms rely significantly more often on informal enforcement mechanisms alone. Unfortunately, there are no separate data available for a distinction of dispute resolution means between customers and producers. But it is expected that the means of dispute resolution used with international partners are more formally organised than those used with Chinese partners. However, this cannot be empirically proven.

The argumentation above has shown that institutions provided by the government play a crucial role in explaining the significantly different results in HK and the PRD. In Table 6.10, it is additionally tested what structural firm characteristics correlate with the use of enforcement techniques used by HK and PRD firms. As some variables could only be measured on a nominal scale (for example HK connection), the contingence coefficient Cramer's V was chosen to measure the strength of association. Values range from 0 (no association) to 1 (the theoretical maximum possible association). The three different enforcement categories with increasing degrees of formalisation are set as dependent variables (negotiation only; negotiation and mediation; negotiation, mediation, arbitration and litigation). The variables tested are given in the far left column (firm size: employees, sales, foundation year, ownership, innovativeness, position in the value chain and HK connection). They can be interpreted as the variables used in the logit models in Section 5.2. Cramer's V value and a short interpretation is also provided.

It is immediately obvious that the correlation between enforcement mechanisms and structural characteristics of firms is low in the PRD. Only firm size in terms of sales is significant. The more sales a firm generates (the larger the firm), the more often firms use the variety of dispute resolution processes available. This confirms that the lack of institutionalised means of dispute resolution seems to have a stronger influence on the choice of alternatives than structural firm characteristics. If the appropriate infrastructure is missing, innovativeness, firm size, ownership or age do not influence the decision on enforcement techniques.

In contrast, in HK structural characteristics seem to be important for the decision on enforcement mechanisms. The ownership structure has a significant influence (Cramer's V = 0.27). The higher the share held by foreigners, the more frequently firms use the entire spectrum of enforcement mechanisms. Although the other influences are not significant, their contingence coefficient is higher absolutely (> 0.20) than for PRD firms. A tendency can be identified. The higher the number of employees and the higher the number of sales, the more often firms prefer formal dispute resolution techniques. Moreover, younger, more innovative firms with a higher position in the value chain tend to use more formal means of settling disputes. As HK's institutional and infrastructural environment encourages the appropriate dispute resolution techniques, HK firms can freely choose between alternatives. Therefore, firm characteristics have a greater influence on the decision than in mainland China, where the institutional environment turns out to be a restrictive force.

To conclude, HK firms tend to tap the full potential of dispute resolution techniques. This is supported by the HK government, which is aiming for HK to become a hub for legal services. They provide the appropriate infrastructure. HK firms combine informal and formal means of dispute resolution to reach the maximum possible output. They first apply informal methods such as negotiation, and if this fails, mediation comes into play. They do not hesitate to go for arbitration and litigation as well. However, the data also showed that almost half of the firms indicated that negotiation was a sufficient means of settling disputes. Informal enforcement mechanisms seem to be a popular choice for settling disputes. In China, however,

Table 6.10.: Enforcement mechanisms and its correlation with structural firm characteristics

Enforcement mechanisms X	PRD survey $n = 218$		HK survey $n = 100$	
	Cramer's V	...the more formal	Cramer's V	...the more formal
Firm size: employees	0.12	no correlation	0.24	the larger ...
Firm size: sales	0.21**	the larger...	0.27	the larger ...
Foundation year	0.17	no correlation	0.21	the younger ...
Ownership	0.07	no correlation	0.27**	the higher foreign shares ...
Innovativeness	0.07	no correlation	0.21	the more innovative ...
Position in VC	0.11	no correlation ...	0.23	the higher the position in the VC ...
HK connection	0.04	no correlation	—	—

Note: Dependent Variable: Enforcement Mechanisms (1 = Negotiation only, 2 = Negotiation and Mediation, 3 = Negotiation, Mediation, Arbitration and Litigation

Source: Calculation based on own survey conducted in DFG SPP 1233 [2007]

informal means of dispute resolution are also frequently used, but more out of necessity than choice, since formal means such as arbitration and litigation are difficult
to realise. In China, the lack of international standards in arbitration and litigation
processes seem to be the restrictive reason for the decision about enforcement mechanisms. Firm characteristics do not function very well as determinants. JOHNSON
et al. (1999) studied contract enforcement in transitional economies: Poland, Ro-
mania, Russia, Slovakia and Ukraine. They also found that informal contractual
methods, such as relational contracting, serve as substitutes for formal mechanisms
at the beginning of the transformation process. But they could also show that an
increasing functionality of courts encourages more firms to consider litigation. On
the other hand, they argue that informal enforcement will retain its importance for
certain issues. They argue that more advanced technology, specific investments and
more goods with subtle quality characteristics make it difficult or impossible for

courts to verify that the contract has been breached. This fits with the findings of HK firms' behaviour.

6.5. SUMMARY AND DISCUSSION OF FINDINGS IN SECTION 6

In this chapter, informality and flexibility in customer producer relations of electronics firms in the GPRD was studied from three perspectives. Firstly, the governance modes and their flexibility were analysed. Secondly, contact channels and selection criteria for customers and producers were analysed. Thirdly, contractual arrangements needing to be settled following the selection process were examined. Additionally, it was mentioned that even if customers and producers are carefully selected and contracts are concluded, disputes can still occur. Informal and formal means of enforcing contracts were analysed. Informality has been studied within its five dimensions outlined in the theoretical part of this work. While analysing formal and informal means, its effect on flexibility was measured. Moreover, a comparison was made between firms localised in HK and those in the PRD to see whether they informalise business practices on different levels.

Flexibility in governance modes was measured in a quantitative and qualitative way. The time of negotiation was an indicator of the quantitative flexibility. The negotiation time increased when moving from hierarchies to markets. Qualitative flexibility was measured by the different options firms consider when order volume suddenly changes. The survey revealed that firms in cooperation and market relationships are provided with more qualitative flexibility to react, whereas firms in hierarchies enjoy more quantitative flexibility. This explains where the competitive edge for each governance mode lies. What could not be proven is that cooperative, more informal modes provide the greatest flexibility. One can still argue over which form of flexibility is or will become more important, but this is not the subject of this work. Cooperation as a medium mode certainly provides for both directions of development.

The analysis of contacting channels (corresponding to the conceptual discussion in Section 2.5.1) of HK firms proved that HK firms use informal modes to recruit customers, even though the institutional environment gives leeway to rely on formal modes alone. Furthermore, the data show that HK firms succeed while combining formal and informal modes. The two extremes are not mutually exclusive, but rather complement each other. Informal modes seem to smooth business relations even in a complete institutional environment. This is in line with the theoretically derived discussion of ZENGER et al. (2002) in Section 2.3.5. Comparing HK and PRD firms, the data show that PRD firms still rely more frequently on informal modes during contact and selection processes. China is still in transition, therefore formal interactions sometimes do not find their basic analogy in the law. Even though the Chinese government has improved the quality of formal channels, firms need time to adapt. They react with a time delay to the introduction of formal means, which has been suggested by WILLIAMSON (2000). Looking at these results, Hypothesis D1—*firms in HK apply informal modes to recruit and select producers and customers, but to a lesser extent than firms in the PRD*—cannot be rejected. Moreover, Hypothesis D2—*the combination of informal and formal practices for recruitment enables firms to be flexible*—also cannot be rejected. A variety of choices ensures finding the most appropriate channel to recruit and select partners. The connection between flexibility and business success could not be proven because of a lack of valid data, but should be the topic of further research.

As soon as customers and producers are selected, contracts come into play (corresponding to the conceptual discussion in Section 2.5.2). It was analysed (1) what share of firms rely on the dynamic and static contracting model, (2) whether the dynamic contracting model reduces time of negotiation and (3) whether informal agreements in particular accelerate the process. The survey showed that half of the firms in HK and the PRD use the dynamic contracting model. The theoretical discussion of MASTEN (2000) and HART and MOORE (1999) about the profitability of incomplete contracts and their renegotiability could be proven as true for the GPRD. The data show that the effect of FAs on the negotiation time in dynamic models depends on the governance modes. When market relationships turn into NEC, the negotiation time nearly halves. In EC and hierarchies, no positive effect could be proven. Therefore, Hypothesis D4—*firms using the dynamic contracting model are more flexible than firms using the static model*—can be confirmed for market relations, but must be rejected for EC and hierarchies.

The informality of FAs in terms of precision and form was also studied. HK firms tend to completely cover the necessary details in FAs, mainly in written form. However, Hypothesis D3—*firms in the GPRD rely on informal contracting modes, but HK firms' contracts with PRD firms are likely to be more informal than contracts with firms in the Western world*—must be partly rejected. Contracts are mainly concluded formally. Although written contracts are adapted to the Western standard, it has been found that firms in China do not rely on a person's word or a handshake anymore. A comparison between customers and producers showed that FAs with producers in the PRD are slightly more informally organised than FAs with international customers. Moreover, it could be proven that firms with informal FAs have a reduced negotiation time in comparison to firms with formal FAs. Hypothesis D5—*informal contracts enhance the flexibility of firms*—can be confirmed: informal contracts lead to a greater quantitative flexibility.

When enforcement of contracts is needed (corresponding to the conceptual discussion in Section 2.5.3), HK firms use the entire spectrum of techniques from informal to formal. Any dispute is likely to be settled by negotiation first. In case of failure, mediation, arbitration and litigation follow. Although about half of HK firms only need to negotiate to settle disputes, they do not hesitate to use legal action as well. The institutional environment supports HK firms' choice to go for the most appropriate dispute resolution technique. In contrast, PRD firms cannot flexibly choose their means of dispute resolution, because national institutions for arbitration and litigation are lacking in quality. Instead, they rely mainly on negotiation. Therefore, Hypothesis D6—*HK firms use the entire spectrum of formal and informal enforcement mechanisms, whereas PRD firms mainly rely on informal techniques*—cannot be rejected. Differences in enforcement mechanisms between HK and PRD firms are significant. Determinants of the use of enforcement mechanisms in the PRD can mainly be reduced to the institutional environment. However, in HK it could be found that larger, younger firms with more foreign shares and better technological skills are more likely to use formal enforcement mechanisms. Hypothesis D7—*firm characteristics determine the choice of enforcement mechanisms*—must be partly rejected. Firm characteristics are only important for HK, and not for PRD firms.

In more general terms, it could be proven that although China is a transitional economy which is continually improving its institutional environment, it firstly has not yet reached the level of HK institutions, and secondly, even if it has reached an international level in some areas, firms have not immediately changed their traditional business practices. According to WILLIAMSON (2000), firms adapt gradually to the new system. Hypothesis B2—*PRD firms use informal practices more intensively than HK firms*—can be confirmed in terms of recruitment and enforcement mechanisms. Contracts in the PRD could not be analysed according to their informality, but it could be shown that HK firms have slightly more informal contracts with producers in the PRD than with international customers.

Hypothesis B3—*although HK firms work within a complete institutional environment, they still rely on informal practices*—can be confirmed. In terms of recruitment of customers and enforcement mechanisms, this hypothesis could be proven. Regarding contracts, HK firms mainly prefer formal contracts fixing the important points. But it was derived from the interviews that formal standards in HK are not in line with those in the Western world. A formal contract to China-based producers leaves more leeway for renegotiation. Hypothesis B4—*informal practices in customer producer relations lead to greater firm flexibility*—can be definitively confirmed for contracts, because informal contracts reduce negotiation time (quantitative flexibility). In recruitment processes and enforcement mechanisms, it could be proven that the use of informal means enhances firms' qualitative flexibility. HK firms in particular benefit from the ability to choose between informal and formal means in order to find the most appropriate solution. Although these findings suggest a direct connection between informality and flexibility, the topic deserves more research in order to provide a holistic picture.

J *How high is the degree of informality in customer producer relations of HK-based and Chinese-based firms? Are there any differences in the behaviour of*

firms between the two locations? Does informal behaviour result in greater
flexibility? In answering these questions, the focus will be on:

- *contact and selection procedures for customer and producers*
- *contractual arrangements*
- *enforcement mechanisms*

HK firms still rely on informal practices to organise their business. However, informal modes are of greater importance in producer relations with the PRD than in customer relations of global scale. PRD firms indicated working even more informally. In contact and selection processes, HK firms combine informal with formal modes. In HK, it could be shown that guanxi has changed from a network of private relations to a network of functional business relations. In the PRD, this development could also be observed, but to a lesser extent. Private relationships are still of greater importance than in HK. In contractual arrangements, the degree of informality is low. HK firms' contracts mainly cover the necessary details in written form, but contracts to PRD producers are slightly more informal. In terms of enforcement mechanisms, negotiation as an informal mode is still frequently used by HK and PRD firms. Negotiation in the PRD mainly functions as a necessity due to an incomplete infrastructure supporting formal modes, whereas in HK, negotiation is an option which can be the most appropriate for settling disputes. In general, using the range from formal to informal practices increases firms' options to react and organise business, thereby increasing qualitative flexibility. Quantitative flexibility resulting from informal interactions could only be proven for contracts in terms of negotiation time. For recruitment and enforcement mechanisms, data about quantitative flexibility were not collected. What could not be proven directly, due to invalid data, is the relationship between flexibility and success. It could be proven that HK firms' ability to predict market changes is low and that PRD firms' ability is even lower. Flexible reactions seem to be a way of enabling firms to deal with market pressure, but it might not be directly related to or measurable in typical business success figures such as sales or growth. But flexible reactions can already be interpreted as a variable of business success itself under certain constraints.

For stronger empirical evidence of the relationship between informality and flexibility, further research is needed which covers other business areas as well. In this work, the focal point was to analyse the informalisation of business practices in customer and producer relations and their effect on flexibility in HK and the PRD.

7. CONCLUSION

The research objective of this work was to analyse how electronics firms in the GPRD are integrated into global value chains in order to respond to the high degree of flexibility required by leading global firms. Emphasis was placed on the firms' selection of governance modes to organise their customer and producer relations, and beyond that, on how informal modes of interactions enhance the flexibility of firms in terms of the contact and selection processes for customers and producers, as well as for the conclusion and enforcement of contracts. Against this background, hypotheses were derived from the conceptual Chapter 2 which were fundamental to answering the empirically guided research questions. This last chapter will consist of an overall summary and discussion of empirical findings (see Section 7.1). This provides the basis for policy recommendations and an outline for further research possibilities (see Section 7.2) as demanded by the policy-related research question **K**.

> **K** *What implications do the research results have for policy makers and firm managers in the GPRD and what further research is needed in order to consolidate the new business concept?*

7.1. SUMMARISING EMPIRICAL FINDINGS

In the first empirical chapter, it could be proven that electronics firms in the GPRD are integrated into modular global value chains. The position in the value chain can differ between firms in HK and the PRD, although this is not always the case. Large HK electronics firms usually take the position of a global contract manufacturer (CM), which serves leading Western firms on an OEM or ODM basis. They organise the entire production process for several leading firms, which allows them to profit from economies of scale. For most SMEs in HK, one can say that they follow the chain and provide CMs both in HK and abroad with components. HK, as a location with a complete institutional setting and a history of trade, provides firms with perfect conditions for doing international business. Two separate surveys of electronic firms were conducted in HK and the PRD in order to follow the organisation of the value chain in the GPRD. The survey of HK SMEs revealed that most manufacturing activities of HK firms take place in the PRD. In order to follow the value chain, a survey of PRD firms was conducted. Almost half of the PRD firms operated as producers for HK firms, while the other half built up relations to international customers. The functionality of HK firms as a gateway to the world market for PRD firms is declining. In any case, the PRD survey revealed that producers of PRD firms are mainly located in the same city or other cities in the PRD. The production process along the value chain in the GPRD is very much locally concentrated. The segment of the electronics value chain in the GPRD starts with large CMs in HK and continues with a network of firms located in the PRD. The spatial distance between leading firms and CMs is large, but the producer network is locally concentrated in the PRD. This spatial dimension has not yet been considered in the value chain approach of GEREFFI et al. (2005), but should be integrated. How these relationships are organised is considered in the second empirical chapter.

The empirical analysis of the applied governance modes to customers and producers within the value chain in the GPRD provided evidence of how those firms are connected. Hierarchies, equity cooperation, non-equity cooperation and markets were distinguished. Equity and non-equity cooperation is related to more informal modes of interaction because it involves trust, personal considerations and reliability. The majority of HK firms preferred a hierarchical relationship to their producers in the PRD, but some firms also opted for equity cooperation or non-equity cooperation. Taking a time dimension into account, data provide evidence that an improving formal institutional environment in the PRD has encouraged more HK firms to opt for cooperative relationships or even market relationships with their PRD producers. The quality of PRD producers has also improved. The "front shop, back factory" model has changed its face. This is because front shop and back factory are no longer controlled by one firm, resulting in an increase in cooperative relationships. A comparison of HK firms' relations to customers and producers revealed that the selection of governance modes (hierarchy → market) depends heavily on the location of partners, which is strongly linked to a particular institutional environment. Relationships to international customers in a complete institutional environment were mainly market-oriented, whereas relationships to producers were still mostly organised in hierarchies, despite recent trends to the contrary. The theoretically derived assumption that the institutional environment determines governance modes could be proven empirically, although there may be arguments that other determinants, such as general business practice, are also partly responsible for this decision. Besides the variable location, which can be decisive, as it is known before the governance mode is selected, it could be shown that relationship-specific characteristics are significantly associated with certain governance modes. The survey showed that firms with a higher dependency on customers or producers and greater working experience were more likely to be organised in hierarchies. For producer relations of HK firms, it could be empirically proven that cooperative and more informal governance modes reduce the time required for negotiation with partners. This increases the flexibility of firms, because the production process is not delayed by extensive contractual negotiations. The organisation within the electronics value chain differs according to the institutional environment. The value chain concept does not take this point into account. Criticism of the value chain concept has touched on this point before, but this study provides empirical evidence to strengthen the call for the expansion of the value chain concept.

In the third empirical chapter, the discussion went beyond the governance modes and the factors associated with them. The level of informality and flexibility in interaction processes with customers and producers was analysed. Special emphasis was placed on contact and selection processes for customers and producers, as well as on the conclusion and enforcement of contracts. The five dimensions of informality served as an analytical framework. The surveys revealed that PRD firms rely more often than HK firms on informal processes while contacting and selecting customers and producers, which is a result of the inadequate support of formal processes by law. The difference in the level of informality can be explained by PRD firms' dependency on informal processes. However, the strategy of HK firms is to use a variety of informal and formal methods to recruit and select customers and producers. A complete institutional environment does not necessarily lead to formal behaviour. Formal and informal processes are not mutually exclusive for HK firms, but they are rather combined to enlarge the pool of opportunities from which firms can choose. Turning to the contractual arrangements, firms in HK and in the PRD mainly fix contracts very precisely in writing. Informal contracts can hardly be found, although PRD firms tend to use more informal elements than HK firms. Contract enforcement in the PRD is more difficult due to a lack of legal safeguards. Therefore, PRD firms rely more frequently on informal mechanisms of conflict resolution, such as negotiation and mediation, while HK firms do not hesitate to use litigation as well. However, informal interactions, such as proximity, face-to-face contacts, mutual help within networks and trust, are more important for PRD managers when doing business. Although the formal institutional environment in China has recently provided more efficient safeguards for contract conclusion and enforcement, Chinese managers need time to adapt their managerial behaviour. Whereas formal institutions can change overnight, informal institutions only change gradually (WILLIAMSON 2000). This explains why firms in the PRD still rely on more informal institutions, even if, in some cases, formal institutions have already been adapted to the international standard. It can be expected that PRD firms will formalise their behaviour to a certain extent, but not completely. Informal and formal interactions seem to work in combination rather than being mutually exclusive, which confirms the expectation of ZENGER et al. (2002). A certain degree of informality will probably be sustained, as could be proven in the case of HK firms. Informal modes of interaction seem to enhance the flexibility of firms, both in terms of speed and variety. Firms benefit from (1) using business contacts to find a new producer or customer quickly, (2) concluding informal contracts to increase speed to market and (3) negotiating rather than litigating, as it is confidential and saves time. It can be concluded that a certain degree of informalisation of behaviour serves the flexibility of firms well. As this is only a rough and preliminary explorative study to test this hypothesis, further research is certainly needed to provide more details of this interplay.

7.2. NEED FOR FURTHER RESEARCH AND IMPLICATIONS FOR POLITICIANS AND MANAGERS

The aim of this work is not only to establish what effects informal modes of interaction in customer and producer relations have on flexibility in the GPRD, but also to encourage and invite further research on the topic of informality. This work investigates only a very limited area of business activities in the GPRD, but the empirical evidence shows that the idea of a new business model driven by informal interactions should be considered. The following recommendations are answers to research question **K**, which asked what further research is needed to consolidate the new business concept:

Further Research Required

Extension of network analysis: To provide a wider scope of the organisation of customer and producer relations, research should be extended to the entire network of customers and producers, not only to the most important partners. This would help to differentiate customer and producer relations with different levels of informality and flexibility.

Transferability to other business operations: The focus of this work was to analyse the degree of informality in external firm organisation with customers and producers. Other external relationships to government officials, innovation partners, universities and business service providers have not been taken into account. Furthermore, the implications of informality for internal firm organisation, for example internal production organisation, administration, financing and innovation activities, has not been considered. For a holistic picture of a business model driven by informality, other business activities must be studied and compared. As this idea is part of a larger research project, two further studies are expected in the future. The first will analyse the informal modes of interactions in innovation activities (Wan-Hsin Liu, The Kiel Institute of the World Economy), while the second study will consider the role of informality in general firm activities from a comparative perspective (Daniel Schiller, Leibniz University of Hannover, Institute of Economic and Cultural Geography).

Considering positive and negative effects of informality: A more intensive analysis of informality in customer producer relations, but also in other business operations, would be valuable in order to better assess the borderline between the positive effects of informality on flexibility and the negative effects, which could lead to a lock-in of firms.

Assessment of the importance of the variable informality for firm flexibility: This work has focussed on the interplay between informality and flexibility. In the introduction, it was mentioned that other influencing factors might exist which also contribute to the flexibility of firms. To assess the importance of informality for firm flexibility, other potentially influencing factors must be considered and their influence compared to that of informality.

Flexibility, informality and firm success: It has been argued that global markets demand flexibility. Operating with a certain degree of flexibility is assumed to increase the success of firms. What has not been analysed in this work is whether a higher degree of flexibility has an effect on the success figures of firms, such as sales or sales growth. If it can be empirically proven that flexibility increases firms' success, then informal behaviour can also be assumed to be related to measurable success figures. It is, therefore, worth studying whether informality has a direct effect on firm performance.

Persistence of informality and upgrading of firms: Informality seems to be a suitable concept when firms are required to serve market demand in manufacturing processes quickly. To keep their competitive edge, firms in the GPRD are forced to upgrade their manufacturing activities and invest in innovation activities. Innovation-related knowledge, partners and contracts are more sensitive. More intensive studies are required to establish whether informality will remain important for firms in upgrading processes. It is expected that trust is more necessary for informal knowledge transfer, but that contracts or patents will become more formalised.

Transferability to other regions: In the GPRD region, there are indications that informal modes of interaction enhance flexibility. It has been argued that this region seems to be suitable for proving this model due to the two different institutional settings. But nevertheless, there is not yet empirical evidence which illustrates to what extent this phenomenon is limited to the GPRD region, or whether the concept is transferable to other regions as well. To conceptualise or theorise the findings of this study, a similar study in other regions is recommended. It is recommended that any such study should (1) concentrate on regions with two different institutional settings (for example Singapore/Malaysia, USA/Mexico), (2) consider other Asian agglomerations in order to establish whether this is a uniquely Asian phenomenon (for example the Shanghai or Beijing regions, the Bangkok region, the Seoul region) and (3) apply the same criteria to Western agglomerations (for example Austria/Hungary, Finland/Estonia).

Transferability to other sectors: This study concentrated on the electronics industry, but the transferability of the phenomenon to other sectors should be investigated. Differences might occur when observing low-technology sectors, such as the textile industry, in comparison to high-technology sectors, such as biotechnology.

Learning effects for German enterprises: In case the concept of informality and flexibility is evident not only for the GPRD region, but also in other regions, one should consider what German enterprises can learn from this business model. The Western business culture seems to be more formalised than the business environment in the GPRD. It would be valuable to analyse what capacity German firms have to adapt to this business model and how this could contribute to their business success.

Although the research focus in this work is limited and additional research is needed, an attempt has been made to derive implications for politicians in HK and the PRD to increase the competitiveness of the region. This also refers to research question **K**. The empirical findings revealed trends in the economic development which should be considered or further emphasised by politicians in the future.

Policy Implications

Definition and concentration on strengths in HK: HK is, in many ways, still the gateway to the PRD, but the capabilities of PRD firms are increasing and they can now compete in certain areas formerly thought of as the sphere of HK firms, for example simple service, innovation activities. HK must clearly define its future role in the GPRD and concentrate on its strengths in order to benefit from the unique setting in the GPRD.

Developing a global service hub in HK: HK should continue to consciously develop itself towards being a global hub of sophisticated services. Legal services, for example, provide potential for upgrading the service activities. In doing so, informal modes of firm behaviour should be considered and taken advantage of. As the value chain becomes more fragmented, a variety of special legal services is needed to conclude (international) contracts. Furthermore, service centres for different dispute resolution techniques should be further developed, for example arbitration and mediation centres. Supporting firms in finding the appropriate method of dispute resolution enhances their competitiveness. HK should establish itself as a service hub not only for firms in the GPRD, but also for firms in the Shanghai or Beijing regions.

Strengthening the interaction between HK and the PRD: HK politicians should
continue to work intensively with regional stakeholders in the PRD. Only cooperation and a development of common strategies can provide the basis for
regional competitiveness. Inconsistent or competing aims waste valuable resources. An organisation should be developed which provides a framework
for interaction between regional stakeholders in HK and the PRD.

Furthermore, the empirical evidence gives rise to recommendations for electronics
firms in the GPRD to optimise and evaluate their business activities. Recommendations for firm managers in HK and the PRD are based on ideas derived mainly from
interviews with firms and regional stakeholders.

Recommendations for Firms

Innovative capitalisation on cross-border business activities: Firms in HK and
the PRD still gain a significant advantage from cross-border business relationships, for example due to better market access, a more intimate knowledge
of global demand, greater predictability of markets or a sophisticated set of
financial and other business-related services. HK and PRD firms should continue to capitalise on the opportunities to combine their resources in order to
successfully operate on markets. Competition should be avoided.

Use informal strategies as an opportunity: Many HK firms view informality as
an inferior business strategy when compared with the highly formalised ways
of doing business in Western economies. However, the findings suggest that
informality is an opportunity to increase flexibility and responsiveness, even
if business environments are improving and formal ways of doing business
are becoming more feasible. PRD firms should follow the HK example of
using a balance of informal and formal modes of behaviour in order to reach
the greatest flexibility.

Improve predictability of changing market conditions: The need to adjust production processes quickly often arises from difficulties in predicting changes
on markets. HK firms have an edge over PRD firms in this respect. HK
firms should interact closely with global customers and transfer forecast information to their PRD producers. Only a sharing of information furthers the
regional competitive advantage.

Transform the organisation of customer producer relations: A wholly-owned
production plant in the PRD is still efficient, but close cooperation with Chinese producers turned out to be a flexible alternative. In the future, firms in
HK should concentrate on their competitive advantage (product quality, marketing, innovativeness) and rely on qualified Chinese producers rather than
operating their own production plant in the PRD.

Combine formal and informal recruitment processes of partners: Firms in HK shall continue to combine formal and informal modes for contacting and selecting customers and producers. Informal modes, such as personal contacts, are a reliable and quick way to select producers. Formal modes take into account objective measurements such as expertise and quality. HK firms are thus recommended to consciously establish an efficient selecting and contacting system for producers to ensure high quality and flexibility in their production processes. PRD firms should learn from HK firms in this respect, as quality measurements are underemphasised in their priority catalogue.

Considering dynamic rather than static contracting models: More firms in HK and the PRD should consider concluding framework agreements in advance in order to reduce the negotiation time when actual orders are placed. This enhances their speed to market since the production process starts without delay. It is more time-consuming to consider all necessary details in a direct contracting process when an order is placed.

Favour appropriate and reliable modes of dispute resolution: HK firms use informal and quick conflict solution mechanisms, such as negotiation, but do not hesitate to use legal action as well when informal mechanisms fail. PRD firms rely mainly on negotiations. They are recommended to consider other forms of contract enforcement if appropriate and feasible.

This work showed that the call of YEUNG (2007) and YEUNG and LIN (2003) for a research agenda for Asian firms corresponds with the empirical findings in the GPRD. Informal modes in interactions enhance the flexibility of firms. Firms in the GPRD use this strategy to compete successfully in the low-cost production of electronic items. Although this work only took into account the organisation of customer producer relations, it can be concluded that the idea of a special business model for firms in Asia must be strengthened. The findings of this work encourage and invite others to follow and further conceptualise the idea. The theoretical approaches and empirical findings for a new concept provided in this work are from the point of view of economic geography. This view should be combined with perspectives from other disciplines in order to reach a comprehensive model of business organisation in Asia which explains the success of firms.

REFERENCES

AGARWAL, R. N. (1999). Financial Liberalisation and Economic Development in China. In: *China Report*, 35, 27–40

AI, J. (2006). Guanxi networks in China: Its importance and future trends. In: *China and World Economy*, 14(5), 105–118

ANGELES, R. and NATH, R. (2000). An empirical study of EDI trading partner selection criteria in customer-supplier relationships. In: *Information & Management*, 37(5), 241–255

ARAUJO, L. and ORNELAS, E. (2007). *Trust-Based Trade*. CEP Discussion Paper No 820, Centre for Economic Performance, London School of Economics and Political Science

AVITTATHUR, B. and SWAMIDASS, P. (2007). Matching plant flexibility and supplier flexibility: Lessons from small suppliers of US manufacturing plants in India. In: *Journal of Operations Management*, 25(3), 717–735

BAKER, W. E. (2000). *Achieving success through social capital: tapping the hidden resources in your personal and business networks*. San Francisco: Jossey-Bass

BALDWIN, R. (2006). *Globalisation: The great unbundling(s)*. Tech. rep., Prime Minister's Office, Economic Council of Finland

BAYUS, B. L. (1998). An analysis of product lifetimes in a technologically dynamic industry. In: *Management Science*, 44(6), 763–775

BERGER, S. and LESTER, R. K. (1997). *Made by Hong Kong*. Hong Kong: Oxford University Press

BICKENBACH, F.; KUMKAR, L. and SOLTWEDEL, R. (1999). *The new institutional economics of antitrust and regulation*. Tech. rep. No 961, Kiel Institute of World Economics

BLUMBERG, B. F. (1998). *Management von Technologiekooperationen - Partnersuche und vertragliche Planung*. Wiesbaden: Gabler Verlag

BOLTON, P. and WHINSTON, M. D. (1993). Incomplete Contracts, Vertical Integration, and Supply Assurance. In: *Review of Economic Studies*, 60(1), 121–148

BURT, R. S. (1992). *Structural holes: the social structure of competition*. Cambridge, Mass.: Harvard Univ. Press

CAI, S. and YANG, Z. (2008). Development of cooperative norms in the buyer-supplier relationship: The Chinese experience. In: *Journal of Chain Management*, 44(1), 55–70

CARNEY, M. (2005). Globalization and the Renewal of Asian Business Networks. In: *Asia Pacific Journal of Management*, 22, 337–354

CENSUS AND STATISTICS DEPARTMENT HONG KONG (n. d.). URL www.censtatd.gov.hk,(10.02. 2009)

CHANG, W. S. and HILMER, S. (2006). China Arbitration Law versus Uncitral Model Law. In: *International Arbitration Law Review*, 9(1), 1–7

CHOI, C. J. (1994). Contract Enforcement across Cultures. In: *Organization Studies*, 15(5), 673–682

CHRISTERSON, B. and LEVER-TRACY, C. (1997). The third China? Emerging industrial districts in rural China. In: *International Journal of Urban and Regional Research*, 21(4), 569–588

COASE, R. H. (1937). The Nature of the Firm. In: *Economica*, 4, 368–405

COASE, R. H. (1992). The Institutional Structure of Production. In: *American Economic Review*, 82(4), 713–719

COLEMAN, J. S. (1988). Social Capital in the Creation of Human-Capital. In: *American Journal of Sociology*, 94, S95–S120

COLEMAN, J. S. (1990). *The Foundations of Social Theory*. Cambridge, Mass.: Belknap Press of Harvard Univ. Press

DE MESQUITA, E. B. and STEPHENSON, M. (2006). Legal institutions and informal networks. In: *Journal of Theoretical Politics*, 18(1), 40–67

DiMAGGIO, P. and LOUCH, H. (1998). Socially embedded consumer transactions: For what kinds of purchases do people most often use networks? In: *American Sociological Review*, 63(5), 619–637

DJANKOV, S.; LA PORTA, R.; LOPEZ-DE SILANES, F. and SHLEIFER, A. (2003). Courts. In: *Quarterly Journal of Economics*, 118(2), 453–517

DONGGUAN MUNICIPAL STATISTICS BUREAU (2007). *Dongguan Statistical Yearbook*. China Statistics Press

DUNNING, J. H. and KIM, C. (2007). The cultural roots of Guanxi: An exploratory study. In: *World Economy*, 30(2), 329–341

DUYSTERS, G.; KOK, G. and VAANDRAGER, M. (1999). Crafting successful strategic technology partnerships. In: *R & D Management*, 29(4), 343–351

ENRIGHT, M.; SCOTT, E. and CHANG, K. (2005). *Regional Powerhouse: The Greater Pearl River Delta and the Rise of China*. Singapore: John Wiley & Sons (Asia)

ENRIGHT, M.; SCOTT, E. and ENRIGHT, SCOTT & ASSOCIATES (2006). *The Greater Pearl River Delta*. Hong Kong: Invest in Hong Kong, 4th edn.

ERNST, D. (2000). *The Economics of Electronics Industry: Competitive Dynamics and Industrial Organization*. Economics Study Area Working Papers No 7, East-West Center

ERNST, D. (2004). Global Production Networks in East Asia's Electronics Industry and Upgrading Prospects in Malaysia. In: S. YUSUF; M. A. ALTAF and K. NABESHIMA (eds.) *Global production networking and technological change in East Asia*, Washington: World Bank

ETGEN, B. (2007). *Unternehmenssteuerreform fuer Unternehmen in der VR China - Auswirkungen auf auslaendische Investoren*. Special Edition April 2007, Beiten Burkhardt Newsletter

FAN, C. C. and SCOTT, A. J. (2003). Industrial agglomeration and development: A survey of spatial economic issues in East Asia and a statistical analysis of Chinese regions. In: *Economic Geography*, 79(3), 295–319

FAN, Y. (2002). Questioning guanxi: definition, classification and implications. In: *International Business Review*, 11, 543–561

FEARON, J. D. and LAITIN, D. D. (1996). Explaining interethnic cooperation. In: *American Political Science Review*, 90(4), 715–735

FEDERATION OF HONG KONG INDUSTRIES (FHKI) (2003). *Made in PRD: The Changing Face of HK Manufacturers*. Tech. rep., Federation of Hong Kong Industries

FEDERATION OF HONG KONG INDUSTRIES (FHKI) (2007). *Made in PRD: Challenges & Opportunities for HK Industry*. Tech. rep., Federation of Hong Kong Industries

FEENSTRA, R. C.; HANSON, G. H. and LIN, S. (2002). *The Value of Information in International Trade: Gains to Oursourcing through Hong Kong*. Tech. rep. 9328, National Bureau of Economic Research

FURUBOTN, E. and RICHTER, R. (1991). The New Institutional Economics: An Assessment. In: E. FURUBOTN and R. RICHTER (eds.) *The new institutional economics*, College Station, Texas: Texas A&M University Press. 1–32

GASTON, R. J. and BELL, S. (1988). *The Informal Supply of Capital*. Tech. rep., Office of Economic Research, U.S. Small Business Administration

GÖBEL, E. (2002). *Neue Institutionenökonomik - Konzeption und betriebswirtschaftliche Anwendung*. Stuttgart: Lucius und Lucius

GEREFFI, G. (2005). The Global Economy: Organization, Governance, and Development. In: N. J. SMELSER and R. SWEDBERG (eds.) *The Handbook of Economic Sociology*, Princeton: Princeton University Press and Russell Sage Foundation. 160–182

GEREFFI, G.; HUMPHREY, J. and STURGEON, T. (2005). The governance of global value chains. In: *Review of International Political Economy*, 12(1), 78–104

GIPOULOUX, F. (2000). Networks and Guanxi: Towards an Informal Integration through Common Business Practices in Greater China. In: C. K. BUN (ed.) *Chinese Business Networks*, Singapore: Nordic Institute of Asian Studies. 57–70

GOFFIN, K.; LEMKE, F. and SZWEJCZEWSKI, M. (2006). An exploratory study of 'close' supplier-manufacturer relationships. In: *Journal of Operations Management*, 24(2), 189–209

GOLDSMITH, J. C.; INGEN-HOUSZ, A. and POINTON, G. H. (2006). *ADR in business: practice and issues across countries and cultures*. Alphen aan den Rijn: Kluwer Law International

GRANOVETTER, M. (1973). Strength of Weak Ties. In: *American Journal of Sociology*, 78(6), 1360–1380

GRANOVETTER, M. (1985). Economic-Action and Social-Structure - the Problem of Embeddedness. In: *American Journal of Sociology*, 91(3), 481–510

GREATER PEARL RIVER DELTA COUNCIL (2006). *The Development of Western Pearl River Delta Region and its Prospects for Collaboration with Hong Kong*. Tech. rep., Greater Pearl River Delta Business Council, Task group on Western Pearl River Delta Development

GUANGDONG PROVINCIAL BUREAU OF STATISTICS (2006). *Guangdong Statistical Yearbook 2006*. China Statistics Press

GUANGDONG PROVINCIAL BUREAU OF STATISTICS (2007). *Guangdong Statistical Yearbook 2007*. China Statistics Press

GUANGZHOU MUNCIPAL STATISTICS BUREAU (GMSB) (2007). *Guangzhou Statistical Yearbook 2007*. China Statistics Press

HAN, S. S. and PANNELL, C. W. (1999). The geography of privatization in China 1978-1996. In: *Economic Geography*, 75(3), 272–296

HART, O. (1988). Incomplete Contracts and the Theory of the Firm. In: *Journal of Law, Economics and Organization*, 4(1), 119–139

HART, O. (2001). Norms and the theory of the firm. In: *University of Pennsylvania Law Review*, 149(6), 1701–1715

HART, O. and MOORE, J. (1999). Foundations of incomplete contracts. In: *Review of Economic Studies*, 66(1), 115–138

HART, O. and MOORE, J. (2004). *Agreeing Now to Agree Later: Contracts that Rule Out but do not Rule*. Harvard Law and Economics Discussion Paper No. 465; Harvard Institute of Econ. Research Disc. Paper 2032

HAYTER, R. and HAN, S. S. (1998). Reflections on China's open policy towards foreign direct investment. In: *Regional Studies*, 32(1), 1–16

HENDERSON, J.; DICKEN, P.; HESS, M.; COE, N. and YEUNG, H. W. C. (2002). Global production networks and the analysis of economic development. In: *Review of International Political Economy*, 9(3), 436–464

HERITAGE FOUNDATION (n. d.). Index of economic freedom. URL www.heritage.org/Index,(10. 02.2009)

HERRIGEL, G. (2007). *Flexibility and Formalization: Rethinking Space and Governance in Corporations and Manufacturing Regions*. Tech. rep., University of Chicago, Department of Political Science

HILMER, S. (2007). Hong Kong's mediation pilot scheme for construction disputes. In: *Australasian Dispute Resolution Journal*, 18(1), 37–43

HITT, M. A.; DACIN, M. T.; LEVITAS, E.; ARREGLE, J. L. and BORZA, A. (2000). Partner selection in emerging and developed market contexts: Resource-based and organizational learning perspectives. In: *Academy of Management Journal*, 43(3), 449–467

HOBDAY, M. (2001). The Electronics Industries of the Asia-Pacific: Exploiting International Production Networks for Economic Development. In: *Asian Pacific Economic Literature*, 15(1), 13–29

HONG KONG EXCHANGES AND CLEARING LTD (HKEx) (n. d.). URL www.hkex.com.hk,(10.02.2009)

HONG KONG GENERAL CHAMBER OF COMMERCE (HKGCC) (2003). *Mainland China / Hong Kong: CEPA Closer Economic Partnership Arrangement*. Tech. rep. CP-0021E

HONG KONG INTERNATIONAL ARBITRATION CENTRE (HKIAC) (n. d.). URL www.hkiac.org, (10.02.2009)

HONG KONG TRADE DEVELOPMENT COUNCIL (HKTDC) (2006a). *Impact of Processing Trade Policy Change on Hong Kong Companies*. Business Alert Issue 11, Hong Kong Trade Development Council

HONG KONG TRADE DEVELOPMENT COUNCIL (HKTDC) (2006b). *Latest Developments in Processing Trade in Guangdong*. Business Alert Issue 12,, Hong Kong Trade Development Council

HONG KONG TRADE DEVELOPMENT COUNCIL (HKTDC) (2007a). *1,800 Electronic Products Subject to Compulsory Eco Labelling*. Business Alert Issue 04, Hong Kong Trade Development Council

HONG KONG TRADE DEVELOPMENT COUNCIL (HKTDC) (2007b). *CEPA V - Opportunities for Hong Kong*. Economic Forum Report, Hong Kong Trade Development Council

HONG KONG TRADE DEVELOPMENT COUNCIL (HKTDC) (2007c). *Cost Escalation and Trends for Export Price Increase - A look at the rising production costs in the PRD*. Economic Forum Report, Hong Kong Trade Development Council

HONG KONG TRADE DEVELOPMENT COUNCIL (HKTDC) (2007d). *New Measures on Processing Trade under Restricted Category Take Effect in August*. Economic Forum Report, Hong Kong Trade Development Council

HONG KONG TRADE DEVELOPMENT COUNCIL RESEARCH DEPARTMENT (HKTDC) (2000a). *China's WTO Accession: Challenges and Opportunities for Hong Kong's Electronics Industry*. Economic Forum Report, Hong Kong Trade Development Council

HONG KONG TRADE DEVELOPMENT COUNCIL RESEARCH DEPARTMENT (HKTDC) (2000b). *Competitiveness and Prospects of Hong Kong's OEM, ODM and Brand Name Business*. Economic Forum Report, Hong Kong Trade Development Council

HONG KONG TRADE DEVELOPMENT COUNCIL RESEARCH DEPARTMENT (HKTDC) (2001). *China's WTO Accession and Implications for Hong Kong*. Economic Forum Report, Hong Kong Trade Development Council

HONG KONG TRADE DEVELOPMENT COUNCIL RESEARCH DEPARTMENT (HKTDC) (2003a). *CEPA: Opportunities for Hong Kong Manufacturing Industries*. Economic Forum Report, Hong Kong Trade Development Council

HONG KONG TRADE DEVELOPMENT COUNCIL RESEARCH DEPARTMENT (HKTDC) (2003b). *From No-names to Brands: Findings of a TDC Survey on Hong Kong's OEM, ODM and OBM Business*. Economic Forum Report, Hong Kong Trade Development Council

HONG KONG TRADE DEVELOPMENT COUNCIL RESEARCH DEPARTMENT (HKTDC) (2004a). *CEPA I & II: Opportunities for Hong Kong*. Opportunities for Hong Kong, Hong Kong Trade Development Council

HONG KONG TRADE DEVELOPMENT COUNCIL RESEARCH DEPARTMENT (HKTDC) (2004b). *Guide to Doing Business in China*. Hong Kong Trade Development Council

HONG KONG TRADE DEVELOPMENT COUNCIL RESEARCH DEPARTMENT (HKTDC) (n. d.). *Hong Kong Directory*. Tech. rep., Hong Kong Trade Development Council. URL http://globalcontactshk.hktdc.com

HOWARD, M. and SQUIRE, B. (2007). Modularization and the impact on supply relationships. In: *International Journal of Operations & Production Management*, 27(11), 1192–1212

HUMPHREY, R. H. and ASHFORTH, B. E. (2000). Buyer-supplier alliances in the automobile industry: how exit-voice strategies influence interpersonal relationships. In: *Journal of Organizational Behavior*, 21(6), 713–730

HUNTER, R. J. and RYAN, L. (2008). A Field Report on the Background and Processes of Privatization in Poland. In: *Global Economy Journal*, 8(1), 1–18

INTERNATIONAL TRADE CENTRE (2001). *Arbitration and alternative dispute resolution: how to settle international business disputes*. Trade law series, Geneva

INVEST IN HONG KONG (2008). How Hong Kong Tackles Financial Crisis. URL www.investhk.gov.hk/UploadFile/IPA_global_financial_crisis.pdf,(02.03.2009)

JOHNSON, S.; MCMILLAN, J. and WOODRUFF, C. (1999). *Contract enforcement in transition*. EBRD, Working Paper No. 45

JOHNSON, S.; MCMILLAN, J. and WOODRUFF, C. (2002). Courts and relational contracts. In: *Journal of Law Economics & Organization*, 18(1), 221–277

JONES, R.; KIERZKOWSKI, H. and LURONG, C. (2005). *What Does Evidence Tell us About Fragmentation and Outsourcing?* HEI Working Papers 09-2004, Paper provided by Economics Section, The Graduate Institute of International Studies

KAIPIA, R.; KORHONEN, H. and HARTIALA, H. (2006). Planning nervousness in a demand supply network: an empirical study. In: *The International Journal of Logistics Management*, 17(1), 95–113

KLEIN, P. G. (2005). The Make-or-Buy Decision: Lessons from Empirical Studies. In: C. MENARD (ed.) *Handbook of new institutional economics*, Dordrecht: Springer

KRAJEWSKI, L.; WEI, J. C. and TANG, L. L. (2005). Responding to schedule changes in build-to-order supply chains. In: *Journal of Operations Management*, 23(5), 452–469

KRUG, B. and HENDRISCHKE, H. (2006a). Framing China: Transformation and Institutional Change. Tech. Rep. ERIM Report Series Reference No-2006-025-ORG

KRUG, B. and HENDRISCHKE, H. (2006b). Institution Building and Change in China. Tech. Rep. ERIM Report Series Reference No-2006-008-ORG

KRUG, B. and HENDRISCHKE, H. (2008). Framing China: Transformation and institutional change through co-evolution. In: *Management and Organization Review*, 4(1), 81–108

LALL, S.; ALBALADEJO, M. and ZHANG, J. (2004). Mapping Fragmentation: Electronics and Automobiles in East Asia and Latin America. In: *Oxford Development Studies*, (32), 407–431

LAVIGNE, M. (1995). *The Economics of Transition. From Socialist Economy to Market Economy*. Basingstoke, Hampshire: Macmillan

LEE, M.-K. (1997). The Flexibility of the Hong Kong Manufacturing Sector. In: *China Information*, 12(1/2), 189–214

LEWIN, A. Y.; LONG, C. P. and CARROLL, T. N. (1999). The coevolution of new organizational forms. In: *Organization Science*, 10(5), 535–550

LI, P. P. (2007). Social tie, social capital, and social behavior: Toward an integrative model of informal exchange. In: *Asia Pacific Journal of Management*, 24, 227–246

LI, S.; PARK, S. H. and LI, S. (2004). The great leap forward: The transition from relation-based governance to rule-based governance. In: *Organizational Dynamics*, 33(1), 63–78

LIEFNER, I. (2006). *Auslaendische Direktinvestitionen und internationaler Wissenstransfer nach China*. Muenster: Lit Verlag

LOH, C. (2002). *Hong Kong SMEs: Nimble and nifty*. CLSA Emerging Markets Investment Bank, 4-2002, EC-HK-001-1, Hong Kong: Civic Exchange

LONG, J. S. (1997). *Regression Models for Categorical and Limited Dependent Variables*. Advanced Quantitative Techniques in the Social Sciences, Thousand Oaks, California: Sage

LONG, J. S. and FREESE, J. (2006). *Regression Models for Categorical Dependent Variables Using Stata*. Texas: Stata Press

LORENZ, E. (1988). Neither Friends nor Strangers: Informal Networks of Subcontracting in French Industry. In: D. GAMBETTA (ed.) *Trust Making and Breaking Cooperative Relations*, New York: Basil Blackwell Ltd. 194–210

LUMMUS, R. R.; DUCLOS, L. K. and VOKURKA, R. J. (2003). Supply Chain Flexibility : Building a New Model. In: *Global Journal of Flexible Systems Management*, 4

LUO, Y. and CHEN, M. (1997). Does guanxi influence firm performance. In: *Asia Pacific Journal of Management*, 14, 1–16

LYONS, B. R. (1994). Contracts and Specific Investment: An Empirical Test of Transaction Cost Theory. In: *Journal of Economics and Management Strategy*, 3(2), 257–278

MACAULAY, S. (1963). Non-Contractual Relations in Business - a Preliminary-Study. In: *American Sociological Review*, 28(1), 55–67

MACLEOD, W. B. (2006). *Reputations, Relationships and the Enforcement of Incomplete Contracts*. Center for Economic Studies & Institute for Economic Research, Working Paper 1730, New York

MACLEOD, W. B. (2007). Reputations, relationships, and contract enforcement. In: *Journal of Economic Literature*, 45(3), 595–628

MARTIN, M. F. (2007). *What's the Difference? Comparing U.S. and Chinese Trade Data*. CRS Report for Congress RS22640, United States Congressional Research Service

MASTEN, E. S. (2000). Contractual Choice. In: B. BOUKAERT and G. DE GEEST (eds.) *Encyclopedia of Law and Economics*, Cheltenham: Edward Elgar

MCMILLAN, J. (1995). Reorganizing Vertical Supply Relationships. In: H. SIEBERT (ed.) *Trends in Business Organization: Do Participation and Cooperation Increase Competitiveness?*, Tuebingen: J.C.B. Mohr. 203–222

MENARD, C. (2004). The economics of hybrid organizations. In: *Journal of Institutional and Theoretical Economics*, 160(3), 345–376

MENARD, C. (2005). A New Institutional Approach to Organization. In: C. MENARD (ed.) *Handbook of new institutional economics*, Dordrecht: Springer

MEYER, S.; SCHILLER, D. and REVILLA DIEZ, J. (2009). The Janus-faced Economy: Hong Kong firms as intermediaries between global customers and local producers in the electronics industry. In: *Tijdschrift voor economische en sociale geografie*, 100(2), 224–235

MILLINGTON, A.; EBERHARDT, M. and WILKINSON, B. (2006). Guanxi and supplier search mechanisms in China. In: *Human Relations*, 59(4), 505–531

MISZTAL, B. (2000). *Informality - Social theory and contemporary practice*. London: Routledge

NATIONAL BUREAU OF STATISTICS OF CHINA (n. d.). URL www.stats.gov.cn/tjbz/hyflbz,(10.02. 2009)

NEE, V. (1992). Organizational Dynamics of Market Transition - Hybrid Forms, Property-Rights, and Mixed Economy in China. In: *Administrative Science Quarterly*, 37(1), 1–27

NOEL, T. (1997). The South-East: The Cutting Edge of China's Economic Reform. In: G. LINGE (ed.) *China's New Spatial Economy*, Hong Kong: Oxford University Press. 72–97

NORTH, D. (1990). *Institutions, institutional change and economic performance*. Cambridge: University Press

NORTH, D. (1997). Transaction Costs Through Time. In: C. MENARD (ed.) *Transaction Cost Economics: recent developments*, Cheltenham: Elgar. 149–160

OKURA, M. (1996). Resource Allocation and Institutional Reforms of China's Banking Sector. In: A. KOHSAKA and K. OHNO (eds.) *Structural Adjustment and Economic Reform: East Asia, Latin America, and Central and Eastern Europe*, Tokyo. 155–186

PAJUNEN, K. and MAUNULA, M. (2008). Internationalisation: A co-evolutionary perspective. In: *Scandinavian Journal of Management*, 24(3), 247–258

PARK, S. H. and LUO, Y. D. (2001). Guanxi and organizational dynamics: Organizational networking in Chinese firms. In: *Strategic Management Journal*, 22(5), 455–477

PENG, M. W. and ZHOU, J. Q. (2005). How Network Strategies and Institutional Transitions Evolve in Asia. In: *Asia Pacific Journal of Management*, 22

PIDDUCK, A. B. (2006). Issues in Supplier Partner Selection. In: *Journal of Enterprise Information Management*, 19(3), 262–276

PIES, I. (2001). Theoretische Grundlagen demokratischer Wirtschafts- und Gesellschaftspolitik - Der Beitrag Oliver Williamsons. In: I. PIES and M. LESCHKE (eds.) *Organisationsökonomik*, Tübingen: Mohr Siebeck. 1–27

POWELL, W. W. (1990). Neither Market nor Hierarchy - Network Forms of Organization. In: *Research in Organizational Behavior*, 12, 295–336

PUTNAM, R. D. (1993). *Making democracy work: civic traditions in modern Italy*. Princeton, NJ: Princeton Univ. Press

PUTNAM, R. D. (2000). *Bowling Alone: The Collapse and Revival of American Community*. New York: Simon & Schuster

REED BUSINESS INFORMATION (n. d.). Top 100 Contract Manufacturers. URL www.edn.com/article/CA6482999.html?text=Top+100+contract+manufacturers,(10.02.2009)

REICHHART, A. and HOLWEG, M. (2007). Creating the customer-responsive supply chain: a reconciliation of concepts. In: *International Journal of Operations & Production Management*, 27(11), 1144–1172

REVILLA DIEZ, J. (1995). *Systemtransformation in Vietnam: Industrieller Strukturwandel und regionalwirtschaftliche Auswirkungen*. Hannoversche Geographische Arbeiten 51, Muenster/Hamburg: Geographische Gesellschaft zu Hannover e.V. and Geographisches Institut der Universitaet Hannover

RICHTER, R. and FURUBOTN, E. (1996). *Neue Institutionenökonomik: Eine Einführung und kritische Würdigung*. Tübingen: Mohr Siebeck

SABEL, C. F. (1993). Studied Trust - Building New Forms of Cooperation in a Volatile Economy. In: *Human Relations*, 46(9), 1133–1170

SAFFU, K. and MAMMAN, A. (2000). Mechanics, problems and contributions of tertiary strategic alliances: the case of 22 Australian universities. In: *Library Consortium Management: An International Journal*, 2(2), 44–53

SANDBORG, D. (2007). *Mediation in Hong Kong: Past, Present and Future*. Conference on: Mediation in Hong Kong: The Way Forward (30 November 2007), Hong Kong: Hong Kong International Arbitration Centre

SAUSSIER, S. (2000). When incomplete contract theory meets transaction cost economics: a test. In: C. MENARD (ed.) *Institutions, contracts and organizations: perspectives from new institutional economics*, Cheltenham: Elgar

SCHAETZL, L. (2000). *Wirtschaftsgeographie 2 - Empirie*. Paderborn: Ferdinand Schoeningh

SCHINDLER, J. W. and BECKETT, D. H. (2005). *Adjusting Chinese Bilateral Trade Data: How Big is China's Trade Surplus?* International Finance Discussion Papers 831, Board of Governors of the Federal Reserve System

SCHUELLER, M. (2002). China nach dem WTO-Beitritt. In: *China Aktuell*, 31(2), 140–150.

SEABRIGHT, P. (2004). *The Company of Strangers: A Natural History of Economic Life.* Princeton: Princeton University Press

SEGAL, I. (1999). Complexity and renegotiation: A foundation for incomplete contracts. In: *Review of Economic Studies*, 66(1), 57–82

SHELANSKI, H. A. and KLEIN, P. G. (1995). Empirical-Research in Transaction Cost Economics - a Review and Assessment. In: *Journal of Law Economics & Organization*, 11(2), 335–361

SHILLER, R. J. and POUND, J. (1989). Survey Evidence on Diffusion of Interest and Information among Investors. In: *Journal of Economic Behavior & Organization*, 12(1), 47–66

SIMON, H. (1961). *Administrative Behavior*, vol. 2nd edition. New York: Macmillan

SINDZINGRE, A. (2006). The Relevance of the Concepts of Formality and Informality: A Theoretical Appraisal. In: B. GUHA-KHASNOBIS; R. KANBUR and E. OSTROM (eds.) *Linking the Formal and Informal Economy: Concepts and Policies*, Oxford: Oxford University Press. 58–74

STEENBERGEN, M. (2008). *Logit and Probit Models.* University of Essex, Essex Summer School in Social Science Data Analysis and Collection, Colchester

STURGEON, T. (2000). *How do we Define Value Chains and Production Networks?* Working Paper 00-010, Massachusetts Institute of Technology - Industrial Performance Center

STURGEON, T. (2002). Modular production networks: a new American model of industrial organization. In: *Industrial and Corporate Change*, 11(3), 451–496

STURGEON, T. (2003). *What Really Goes on in Silicon Valley? Spatial Clustering and Dispersal in Modular Production Networks.* Working Paper 03-001, Massachusetts Institute of Technology - Industrial Performance Center

STURGEON, T. (2006a). The Changing Role of the Guadalajara (Jalisco State) Mexico - Electronics Cluster in a Modular Global Value Chain. Conference: Industrial Upgrading, Offshore Production, and Labor at Social Science Research Institute, Duke University

STURGEON, T. (2006b). *Modular Production's Impact on Japan's Electronics Industry.* Working Paper 06-001, Massachusetts Institute of Technology - Industrial Performance Center

STURGEON, T. (2007). Conceptualizing Integrative Trade: The Global Value Chains Framework. In: T. STURGEON (ed.) *Trade Policy Research 2006*, Ottawa: Foreign Affairs and International Trade Canada. 35–72

SULL, D. N. (2005). *Made in China : what Western managers can learn from trailblazing Chinese entrepreneurs.* Boston: Harvard Business School Press

TAYLOR, C. R. (2000). The old-boy network and the young-gun effect. In: *International Economic Review*, 41(4), 871–891

THE GREATER PEARL RIVER DELTA BUSINESS COUNCIL (2007). *Implications of Mainland Processing Trade Policy on Hong Kong.* Research Report, Hong Kong

TIROLE, J. (1999). Incomplete contracts: Where do we stand? In: *Econometrica*, 67(4), 741–781

TRANSPARENCY INTERNATIONAL (n. d.). Transparency International Corruption Perception Index (CPI). URL www.transparency.org/policy_research/surveys_indices/cpi,(10.02.2009)

UNITED NATIONS COMMISSION ON INTERNATIONAL TRADE LAW (UNCITRAL) (n. d.). URL www.uncitral.org,(10.02.2009)

UNITED NATIONS COMMODITY TRADE STATISTICS DATABASE (n. d.). URL www.comtrade.un.org,(10.02.2009)

UNITED NATIONS STATISTICS DEVISION (n. d.a).   URL http://unstats.un.org/unsd/cr/registry/
    regcst.asp?Cl=2,(10.02.2009)
UNITED NATIONS STATISTICS DEVISION (n. d.b). URL http://unstats.un.org/unsd/cr/registry/regct.
    asp,(10.02.2009)
UZZI, B. (1996). The sources and consequences of embeddedness for the economic performance of
    organizations: The network effect. In: *American Sociological Review*, 61(4), 674–698
VOLBERDA, H. W. (1996). Toward the flexible form: How to remain vital in hypercompetitive
    environments. In: *Organization Science*, 7(4), 359–374
WALLIS, J. J. and NORTH, D. (1986). Measuring the Transaction Sector in the American Economy,
    1870-1970. In: S. ENGERMAN and R. GALLMAN (eds.) *Income and Wealth: Long-Term
    Factors in American Economic Growth*, Chicago: University of Chicago Press
WANG, Y. and NICHOLAS, S. (2007). The formation and evolution of non-equity stragetic alliances
    in China. In: *Asia Pacific Journal of Management*, 24, 131–150
WEI, Y.; LIU, X. and LIU, B. (2004). *Entry Modes of Foreign Direct Investment in China: A
    Multinomial Logit Approach*. Working Paper 2004/001, Lancaster: Lancaster University Man-
    agement School
WHITFORD, J. and POTTER, C. (2007). The State of the Art - Regional economies, open networks
    and the spatial fragmentation of production. In: *Socio-Economic Review*, 1–30
WILLIAMSON, O. (1979). Transaction-Cost Economics - Governance of Contractual Relations. In:
    Journal of Law & Economics, 22(2), 233–261
WILLIAMSON, O. (1987). *The economic Institutions of Capitalism: firms, markets, relational con-
    tracting*. New York: Free Press
WILLIAMSON, O. (1991). Comparative Economic-Organization - the Analysis of Discrete Struc-
    tural Alternatives. In: *Administrative Science Quarterly*, 36(2), 269–296
WILLIAMSON, O. (1998). Transaction cost economics: How it works; Where it is headed. In:
    Economist, 146(1), 23–58
WILLIAMSON, O. (2000). The new institutional economics: Taking Stock, Looking Ahead. In:
    Journal of Economic Literature, 38(3), 595–613
WILLIAMSON, O. (2005). Transaction Cost Economics. In: C. MENARD (ed.) *Handbook of new
    institutional economics*, Dordrecht: Springer
WORLD CUSTOMS ORGANIZATION (n. d.).   URL www.wcoomd.org/home_wco_topics_
    hsoverviewboxes_tools_and_instruments_hsnomenclaturetable2002.htm,(10.02.2009)
WORLDBANK (n. d.). Doing Business. URL www.doingbusiness.org,(10.02.2009)
XIN, K. R. and PEARCE, J. L. (1996). Guanxi: Connections as substitutes for formal institutional
    support. In: *Academy of Management Journal*, 39(6), 1641–1658
YANG, C. (2007). Divergent hybrid capitalisms in China: Hong kong and Taiwanese electronics
    clusters in Dongguan. In: *Economic Geography*, 83(4), 395–420
YEUNG, G. (2002). WTO accession, the changing competitiveness of foreign-financed firms and
    regional development in Guangdong of Southern China. In: *Regional Studies*, 36(6), 627–642
YEUNG, H. W. C. (2007). Unpacking the business of Asian business. In: H. W.-C. YEUNG (ed.)
    Handbook of Research on Asian Business, Cheltenham: Edward Elgar. 1–16
YEUNG, H. W. C. and LIN, G. C. S. (2003). Theorizing economic geographies of Asia. In:
    Economic Geography, 79(2), 107–128
YU, T. F.-L. (2000). Hong Kong's entrepreneurship: behaviours and determinants. In: *En-
    trepreneurship & Regional Development*, 12, 179–194
ZENGER, T. R.; LAZZARINI, S. G. and POPPO, L. (2002). Informal and formal organization in
    new institutional economics. In: *New Institutionalism in Strategic Management*, 19, 277–305

214 References

ZHANG, Y. and ZHANG, Z. G. (2006). Guanxi and organizational dynamics in China: a link between individual and organizational levels. In: *Journal of Business Ethics*, 67(4), 375–392

ZHOU, L. (2006). *China Business - Environment, Momentum, Stategies, Prospects*. Singapore: Prentice Hall

APPENDIX A. DEFINITION OF ELECTRONICS INDUSTRY ACCORDING TO STATISTICAL STANDARDS

Table A.1.: Definition of product classification in the electronics industry according to statistical standards

Level	Name	Digits
International and Hong Kong	(HK)-SITC Rev. 3[a]	• 75 Office machines and automatic data-processing machines • 76 Telecommunications and sound-recording and reproducing apparatus and equipment • 77 Electrical machinery, apparatus and appliances • 87 Professional, scientific and controlling instruments and apparatus • 88 Photographic apparatus, equipment and supplies and optical goods
China	HS 2002[b]	• 84 Nuclear reactors, boilers, machinery and mechanical appliances • 85 Electrical machinery and equipment and parts thereof; sound recorders and reproducers, television image and sound recorders and reproducers, and parts and accessories of such articles • 90 Optical, photographic, cinematographic, measuring, checking, precision, medical or surgical instruments and apparatus; parts and accessories thereof

[a]Hong Kong Standard International Trade Classification
[b]Harmonised System

Source: UNITED NATIONS STATISTICS DEVISION, WORLD CUSTOMS ORGANIZATION

Table A.2.: Definition of activity classification in the electronics industry according to statistical standards

Level	Name	Digits
Inter-national	ISIC Rev. 3[a]	• 30 Manufacture of office, accounting and computing machinery • 31 Manufacture of electrical machinery and apparatus n.e.c. • 32 Manufacture of radio, television and communication equipment and apparatus • 33 Manufacture of medical, precision and optical instruments, watches and clocks
Hong Kong	HSIC Ver. 1.1[b]	• 383 Radio, television and communication equipment and apparatus • 384 Electronic parts and components • 385 Electrical appliances and houseware and electronic toys • 386-387 Machinery, equipment, apparatus, parts and components • 389 Professional scientific, measuring controlling equipment
China	GB/-T4754-2002	• 39 Manufacture of electrical machinery and equipment • 40 Manufacture of communication equipment, computers and other electronic equipment • 41 Manufacture of instruments, meters and machinery for cultural and office use

[a]International Standard Industrial Classification
[b]Hong Kong Standard Industrial Classification

Source: UNITED NATIONS STATISTICS DEVISION, CENSUS AND STATISTICS DEPARTMENT HONG KONG, NATIONAL BUREAU OF STATISTICS OF CHINA

APPENDIX B. TEST OF MULTICOLLINEARITY
FOR GENERALISED ORDERED LOGIT MODELS

Table B.1.: Pearson correlation coefficient of variables used in GOL (1)

	(1)	(2)	(3)	(4)	(5)	(6)	(7)	(8)
(1) Location China	1							
(2) Working exper.	0.153	1						
(3) Dependency	0.482	0.237	1					
(4) Time negotiation	−0.246	−0.093	−0.127	1				
(5) Predictability	−0.038	−0.002	−0.039	0.101	1			
(6) Unit in HK	−0.359	0.050	−0.204	0.132	−0.061	1		
(7) Unit family led	−0.020	−0.063	−0.079	0.033	0.014	0.067	1	
(8) Unit linked	−0.041	−0.177	−0.030	−0.045	0.019	−0.031	−0.208	1

Source: Calculation based on own survey conducted in DFG SPP 1233 [2007]

Table B.2.: Pearson correlation coefficient of variables used in GOL (2)

	(1)	(2)	(3)	(4)	(5)	(6)	(7)	(8)	(9)	(10)
(1) Core PRD	1									
(2) Working experience	0.017	1								
(3) Time for negotiation	0.081	−0.294	1							
(4) Innovativeness	0.008	0.182	−0.036	1						
(5) Dependency	0.159	0.253	−0.256	0.280	1					
(6) Predictability of producer information	0.090	0.020	0.036	−0.145	0.101	1				
(7) Position in VC	−0.010	−0.013	0.030	0.302	−0.039	−0.049	1			
(8) Employees	−0.007	0.350	−0.118	0.187	0.191	0.011	0.003	1		
(9) Foundation year	−0.059	−0.605	0.040	0.029	0.053	0.051	−0.005	−0.437	1	
(10) Sales growth	0.039	−0.224	0.023	0.139	0.191	0.094	0.085	0.106	0.202	1

Source: Calculation based on own survey conducted in DFG SPP 1233 [2007]

APPENDIX C. TEST OF THE PARALLEL
REGRESSION ASSUMPTION

Table C.1.: Brant-Test for ordered logit models

	χ^2	$p > \chi^2$	df
All	74.36	0	20
Core PRD	3.96	0.138	2
Working experience	0.99	0.610	2
Time for negotiation	1.75	0.417	2
Innovativeness	1.53	0.465	2
Dependency	4.75	0.093[a]	2
Predictability of producer information	0.45	0.797	2
Position in VC	0.93	0.628	2
Employees	4.13	0.127	2
Foundation year	3.73	0.155	2
Sales growth	0.37	0.832	2

[a]A significant test statistic provides evidence that the parallel regression assumption has been violated

Source: Calculation based on own survey conducted in DFG SPP 1233 [2007]

Geographische Risikoforschung

Heike Egner / Andreas Pott (Hg.)

Geographische Risikoforschung

Zur Konstruktion verräumlichter Risiken und Sicherheiten

Erdkundliches Wissen – Band 147

2010. 242 Seiten mit
18 Abbildungen und
3 Tabellen. Kart.
ISBN 978-3-515-09427-6

Risiken haben Konjunktur. Ob Behörden, Versicherungen, Massenmedien, Stadtentwickler(innen) oder Risikoforscher(innen): Sie alle identifizieren Risiken, warnen vor ihnen und stellen Sicherheiten in Aussicht oder in Frage. Dazu gehört der Hinweis, dass diese ungleich verteilt sind. Zwecks Risikomanagements werden sie verortet und kartiert, sichere von unsicheren Räumen unterschieden.

Der Band plädiert für einen Wechsel des üblichen Beobachtungsmodus: Statt Risiken festzustellen und zu verräumlichen, geht es qua Beobachtung 2. Ordnung um die Praktiken der Konstruktion: Wie und unter welchen Bedingungen werden Risiken konstruiert? Welche Folgen hat ihre Verortung? Im gemeinsamen Rekurs auf die Beobachtungstheorie vereint der Band human- und physiogeographische Perspektiven. Mit seinen Beiträgen demonstriert er die Fruchtbarkeit dieses Ansatzes für die interdisziplinäre Risikoforschung. Als *working group book* lotet er zugleich die Potenziale und Grenzen einer (radikal) konstruktivistischen Perspektive aus.

FRANZ STEINER VERLAG
Postfach 101061 · D-70009 Stuttgart
www.steiner-verlag.de · service@steiner-verlag.de
Telefon: 0711 / 2582-0 · Fax: 0711 / 2582-390